ROUTLEDGE LIBRARY EDITIONS:
FORESTRY

Volume 6

THE MANAGEMENT
OF FORESTS

THE MANAGEMENT
OF FORESTS

F. C. OSMASTON

Routledge
Taylor & Francis Group

LONDON AND NEW YORK

First published in 1968 by George Allen & Unwin Ltd.

This edition first published in 2024
by Routledge
4 Park Square, Milton Park, Abingdon, Oxon OX14 4RN

and by Routledge
605 Third Avenue, New York, NY 10158

Routledge is an imprint of the Taylor & Francis Group, an informa business

British Library Cataloguing in Publication Data
A catalogue record for this book is available from the British Library

ISBN: 978-1-032-77116-8 (Set)
ISBN: 978-1-032-77104-5 (Volume 6) (hbk)
ISBN: 978-1-032-77114-4 (Volume 6) (pbk)
ISBN: 978-1-003-48127-0 (Volume 6) (ebk)

DOI: 10.4324/9781003481270

Publisher's Note
The publisher has gone to great lengths to ensure the quality of this reprint but points out that some imperfections in the original copies may be apparent.

Disclaimer
The publisher has made every effort to trace copyright holders and would welcome correspondence from those they have been unable to trace.

THE MANAGEMENT OF FORESTS

by

F. C. OSMASTON

Commonwealth Forestry Institute, Oxford

London
GEORGE ALLEN AND UNWIN LTD
RUSKIN HOUSE   MUSEUM STREET

PRINTED IN GREAT BRITAIN
in 11 *on* 12*pt. Times type*
BY ABERDEEN UNIVERSITY PRESS

PREFACE

This book has been written primarily for students of forestry, particularly those studying for a University degree. I have, therefore, tried to present the principles and methods on which forest management has been founded in the past and how those principles and methods still apply or are affected by modern conditions which are producing both new aids to management and the complications of new knowledge and skills, changing demands, social habits and resources. It is also to help students and others who may be compiling their first working plan that the chapter on that subject is comparatively long.

The book, therefore, is concerned mainly with the technical aspects of forestry, what may be termed the classic foundations of management and only deals very shortly with the factors of commercial efficiency, labour relations and the many technical aspects of business management and economy, but it discusses some of the implications that arise. The historical chapter at the end tries to show the trends in the development of forestry, particularly in Europe; to provide some guide to the inevitable phases which occur in management and thereby to help the forester to forecast the next phase and decide the appropriate courses of action.

I apologise for repetition and re-elaboration of many matters. This I have done not only for additional emphasis but to make separate parts of the book more complete in themselves and so avoid the need for frequent back references.

I hope the book will also have some value for professional foresters, partly for reference and partly to stimulate a critical study of attractive new methods or practices which may contravene some of the principles and methods stoutly advocated in the past. Although the book naturally deals mainly with practices and conditions in Europe, particularly in Britain, I hope that the experience of success and failure there will help foresters elsewhere.

I am indebted to two Professors of Forestry at Oxford, Professor Sir Harry Champion and Professor M. V. Laurie, and to all my colleagues on the staff of the Commonwealth Forestry Institute for their influence on me during the last thirteen years in exchange of opinion in class and conversation, which have widened yet sharpened my own. In particular, I thank Dr Michael Gane and Dr Eustace W. Jones for reading and criticizing Chapters IV and X respectively; any errors still found are my own.

I am also indebted to several publishers and others for their kind permission to reproduce their tables and figures in the text and Appendices as shown below:

1. Her Majesty's Stationery Office, London. Extracts from Forestry Commission Booklet No. 16: Thinning control tables, Production forecast tables and Normal Yield Tables in Appendix II (a), (b) and (c).

2. The Forestry Commission of Great Britain. Forms extracted from their Working Plans Code, in Appendices I and III.

3. The Food and Agriculture Organization of the United Nations. Extracts from 'Wood; World Trends and Prospects, Unasylva 20 (1–2)' in Tables I, II, III and IV.

4. Verlag Sauerländer & Co., Aarau. Table taken from their book, Hermann Knuchel (1950), *Planung und Kontrolle im Forstbetrieb*, in Table 17.

5. Oliver and Boyd, Ltd, Edinburgh. Tables taken from their book, H. Knuchel, *Planning and Control in the Managed Forest*, translated by Mark L. Anderson (1953), First English Edition, in Tables 10 and 16.

6. The Hildreth Press, Bristol, Conn. Table taken from their book, Herbert H. Chapman (1950), *Forest Management*, 2nd edition, in Table 9.

7. Monsieur L. A. Favre, from personal correspondence on figures and meteorological results at Couvet, in Chapter VIII.

St Cross College
Oxford   F. C. OSMASTON

CONTENTS

1*

APPENDICES

Chapter I

THE SCOPE OF
FOREST MANAGEMENT

DEFINITION OF MANAGEMENT

The management of any enterprise or undertaking clearly includes the organization and conduct of all operations that are needed to fulfil the purpose of the owner or owning body. But that straightforward definition excludes or at least avoids any mention of policy which constantly mingles with the purpose and conduct of any business. The owner may have a single, undivided purpose such as the attainment of the greatest financial profit, but his policy on the course of action to take to win that profit, remains to be settled; he must choose what the main activities and products of his enterprise shall be, where those activities shall be located and what the basic means of producing the products shall be with the skills and resources at his command. For example, the purpose of an owner may be to obtain the utmost profit from his land. Policy must decide which parts of his land should be allotted to suitably chosen activities—e.g. to agriculture, forestry, buildings, roads or other activity. Policy may be taken further still by choosing the type of agriculture to pursue—arable or pasture, the type of forestry, shelter belts perhaps or coniferous pulpwood. Choice and location of the main products to be produced are elements of policy within the owner's basic purpose and become locally the objects of management.

The chief manager, managing director, or chairman of a board of directors of an enterprise is responsible to the owner both for advice on the purposes of an undertaking and their distribution and also for decisions on policy, i.e. the choice of the basic means of attaining the purpose. The chief manager will also naturally be responsible for supervising and regulating the conduct of the whole affair. The lesser managers will be responsible for implementing the policy by planning an organization and conducting the operations. All, even the local manager in the lowest spheres of management, are constantly faced with problems of deciding how a thing should be done, problems made easier by understanding why it should or should not be done and by using a mental or written plan of work. Management is always involved in choosing between two or more courses of possible action.

Therefore in its higher levels at least, management is inextricably entwined with decisions of policy so that the definition of management must be extended to include it. We may conclude that the management of any enterprise embraces three main, interlinked functions, namely settlement of the purposes and main policy to be pursued, the consequent planning and organization of activities, and the conduct of operations.

Policy

Decisions on policy require an assessment of the purposes of the owners in relation to the resources available and the future prospects and conditions foreseen so that the basic means of applying the resources can be decided and defined as a policy. That assessment will demand a clear understanding of several factors, namely:

(i) The physical resources, i.e. the interaction of land, labour and capital, and their influence on the proposed functions and scope of the enterprise.

(ii) The technical foundations and working of the operations entailed.

(iii) The trend of demand on the potential products of the enterprise.

(iv) The human aspects involved in executing all operations.

Any of those factors may be unstable. Policy and even definitions of purpose or at least its location are, therefore, liable to change and should be under constant review. But when revised the purpose must be clearly re-defined with an order of priorities. Confusion of purpose confuses action.

Planning and organization

This branch of management decides how the resources of the enterprise should best be used to attain its purpose and policy.

The planning of how an enterprise will function may be said to be fivefold:

(i) Provision and lay-out of the productive machinery needed to create the desired products, whether the products are material goods or services. This will form the basic capital of the enterprise—in machines, buildings and plant designed, erected and arranged in the most productive manner.

(ii) Arrangement for supplies of materials best suited to drive and feed the productive machinery—e.g. power and unprocessed or raw material for manufacture of the desired material or service product.

(iii) Arrangements for the disposal of the products of the enterprise to consumers—by sale, gift or exchange as befits the purpose of the enterprise.

(iv) Organization of personnel to conduct the enterprise as a whole by allocations of duties and responsibilities.

(v) Finance.

Conduct and control

This aspect of management concerns the daily running of an enterprise and the practical use of its designed structure. It might be called the administration of the enterprise.

Besides mechanical aspects such as maintenance and repair of plant, and financial aspects such as accountancy, the conduct of an undertaking is mainly one of human relations so that the best results are obtained from the planned organization only by harmonious use of the whole man-power whether it is an element in the management itself or in an executive job. This daily direction of man-power demands from management perhaps more art than science, though many principles inherent in leadership can be scientifically formulated, and constitutes perhaps the greatest single element of success or failure.

But all those aspects of management are closely interdependent and each is essential for success. Planning is impossible without a defined purpose and policy; a structure perfectly designed for a purpose is only as good as the conduct and control of it; the most willing and able man-power cannot succeed with poor tools and conversely good tools are useless in the hands of unskilled or unwilling man-power. Nor can policy making or planning ever stop and allow administration to take over on its own. If it did stop the working of an undertaking would become static and fail to develop or adapt itself to the inevitable change of conditions. Without development an undertaking will wither and die. So guided by experience, by technical and inventive progress, by changes in consumer demands and other conditions, policy decisions and planning must develop continuously and harmoniously with the daily administration.

THE MANAGEMENT OF FORESTS

Forest management is no exception to those observations on general principles of management. Forestry does, however, have certain peculiarities that distinguish it from other forms of industry, other sources of raw material and even other uses of land. Those distinctive peculiarities naturally affect the management of forests and it is necessary to examine them briefly.

There are four main peculiarities inherent in forestry, namely: the length of time taken to make its products, the identity of the manufacturing plant and the end product itself, the multiple and varied uses of forests and the extent, topography and accessibility of land used.

The time factor

The growth of trees to useful sizes is slow. So the time that elapses between the formation or origin of a stand of trees and its final harvesting and replacement is long—seldom less than ten years, frequently more than fifty years and not infrequently more than 100 years or even as much as 200 years.

If, however, a forest is grown to provide a service—recreation or protection against wind, for example—there is still a time-lag between the creation of the forest and full realization of its benefits. But once the forest has been satisfactorily grown to an adequate age enjoyment of the service may be continuous from the whole area without intervening periods of waiting for new growth to supply the desired product and perhaps also without elaborate or heavy tending or regeneration expenses. The time factor, therefore, has a much smaller influence in forests grown to provide a service than in those grown for material goods.

Allied with this long wait between formation and harvest of material goods is the fact that different species of trees and even the same species on different sites grow at different speeds. Moreover trees at different stages of growth can provide different usable products—fencing poles in youth, constructional or veneer timber at a mature age. Those two peculiarities impose two problems on the forest manager, how to combine the management of trees growing at different rates in one and the same forest and how to choose between different types of product.

But there are two basic problems caused by the time factor. Firstly, markets may change during the period of growth and thereby invalidate the original decision on what product to grow. An example still affects many woods in England. Until 100 years ago oak was grown in large quantities mainly for naval, but also for constructional timber. Its value for those purposes has gone and 100 years or more spent in growing such trees is largely wasted. Another example is the coppice woods of Europe, now largely useless owing to substitutes for wood fuel, fencing material and so on. Change from growing one kind of forest produce to another is not only expensive but necessarily slow. The need for great care in deciding what to grow is obvious and requires most careful and difficult forecasts of demand which must be kept under continuous review.

Secondly, the long production period involves delayed returns from the initial investment so that much capital is tied to the volume of maturing timber on the ground. This inevitable delay in obtaining returns from capital invested requires special thinking in any forest enterprise.

Identity of product and manufacturing plant

The raw material, wood, produced by forests is formed as a thin annual layer round the stems and branches of trees, the result of circumferential growth dependent on the absorption of water and nutrients from the soil, of oxygen from the air and on photosynthesis by the leaves which use carbon dioxide of the air and energy from heat and light. So the trees themselves are the machinery which manufacture the products desired and direct solar energy is the power to drive the machinery which is fed by the raw materials of water and soil nutrients.

Given a continuance of the right conditions forests are perpetual producers of wooden raw material. Unlike the supplies of mineral produce, supplies of forest raw material can be inexhaustible. That is the great advantage that forests share with other living producers of raw material. But the fact that the stock of trees is both the manufacturing plant and the consumable produce, in other words both the capital employed and the dividend or interest earned, is a complication which is enhanced by the long production period.

It is not possible to harvest the product separately from the capital—only whole trees (only very occasionally branches) when old or big enough to be usable, can be harvested. In consequence two problems are imposed. Firstly, the quantity of whole trees cut must be related to the growth or increment of the whole consortium of trees so that the quantity of trees left after cutting remains sufficient to supply the capital or machinery considered expedient for continuous growth. Secondly, the whole trees cut must suit the customer and the purpose of management (i.e. be the type of produce wanted) and must also be so located that their harvesting is facilitated and the health of those left is furthered.

Clearly this similarity or even identity in form of capital and income, combined with the length of the production period must require a special approach to the management of forests. But the peculiarity of these factors has confused even a legal understanding of them. Thus the general law of England has failed to understand or define the necessary attitude to capital and income in forestry (Gordon, 1965) so that the management of leased and settled forest estates has been made difficult or impossible unless special steps are taken in wording the deeds.

Multiple use

Forests satisfy a great range of human needs which vary from tangible raw materials to intangible benefits. The raw materials have numerous forms—pea sticks to great veneer logs, pulp wood to gums or medicinal herbs while the intangible products vary from aesthetic enjoyment to protection from avalanche.

This multiplicity of product is common to many under-takings other than forestry and is not an unusual complication in management. The peculiarity in forestry is related to the difficulty in deciding the priority to be given to the production of each benefit when several benefits, not all assessable in terms of money, can be provided from the same piece of land. The same piece of land may be unable to provide the maximum value of each benefit. Thus aesthetic enjoyment or protection from avalanches usually requires little or no felling of trees so that economic production of timber is prevented or greatly reduced. It may, however, be possible to segregate the production of each main product in separate areas. Management is then greatly facilitated unless the variation of site qualities both affects seriously what can best be grown and forms a complicated spatial pattern.

Nevertheless it is the variety of continuously replaceable products, tempered, however, by the length of the production period, that gives forests their great value. Not only can a great number of people enjoy them aesthetically, in recreation or in scientific study, but many industries and much employment depend on the raw materials they produce. Additionally their protective values—against wind, erosion or avalanche, and by regulation of water supplies—may make some forests essential however much they cost to run. But those intangible products, frequently intermingled with the production of tangible products in one and the same forest, make it difficult to assess and justify the management of some forests in terms of money. The difficulty of decision is greatest in public forests where it is hard to assess fairly the claims and needs of multiple beneficiaries. To what extent, for example, should recreation be subsidized by reduced profits from growing hardwood instead of coniferous timber? Justification of having and managing forests may not, therefore, be easy—until they disappear when their loss proves their past value.

Extent, topography and accessibility of forest land

Agriculture necessarily occupies the land that is most suitable for growing food crops and it is among those lands that populations have settled. So agricultural land is generally accessible, fertile and comparatively flat. Forests consequently often occupy the more remote, less accessible, less fertile and un-ploughable lands so that they are usually found, often in large tracts, in the mountains and hills or on unfertile soils in the plains. Other forests in comparatively small bits and pieces, occupy land surrounded by agriculture—along steep-sided valleys, on rocky hillocks or as protective belts or groups.

The consequent problems imposed on forest management are those of dispersal of effort and personnel and the difficulties of supervision and protection of the property. The problems and expense of communications and difficulties of extracting timber may be very great.

Technical factors

A basic feature of forest management is the fact that forestry like agriculture is a primary use of the land surface. Consequently forest management requires an understanding of the factors that influence the growth of vegetation and, in particular, tree vegetation. Those factors are clearly complex as they involve not only the botany and physiology of plants and their ecology, but also geology and pedology and the effects of soils and climate on growth. The forest manager must be able to apply this understanding so that he can make the land produce the greatest growth of the desired produce successively from year to year.

But the forest manager is responsible not only for ensuring satisfactory growth of plants in particular trees, but also for their harvesting and sale as well as their replacemnt with seedling trees. That tending, harvesting and replacement require road and other communications, buildings, mechanical equipment and the like.

Therefore the technical ramifications of forestry are wide, ranging from botanic taxonomy, anatomy and physiology to engineering, from land and vegetational survey to soil science, from mensuration of quantities and growth of trees to sociology.

The wide range of skills involved in forestry is outlined opposite.

Naturally no one person can be expert in that array of skills. Yet the forest manager must be able to use them and plan to co-ordinate action for the best results.

Similarly no one book on forest management can attempt to include all the subject matter relevant to forest management as a whole or relevant to the daily working of forests. This book will, therefore, deal with the principles inherent in planning and organization of continuous forest operations—the technical foundations of forest management—and those matters essential to planning decisions and not those matters of administration which maintain or improve efficiency in the execution of a plan. Nor will technological aspects either within the forest estate, such as logging or road construction, or without the forest, such as sawmills, timber preservation or other ancillary enterprises, be included except in as far as they influence the planning of operations in the forest estate.

FOREST MANAGEMENT

Biological Sciences	*Sociological Sciences*
Plant morphology	Economics
Plant physiology	Finance and accountancy
Plant taxonomy	Marketing
Ecology	Social and labour relations
Geology	Law
Pedology	Work study
Hydrology	
Entomology	*Land Use*
Plant pathology	Grazing and pasture
Silviculture	Recreation
Wood anatomy	Wild life
Genetics	Landscape design
	Water supplies
Physical Sciences	
Engineering	*Technology*
Survey and mensuration	Logging
Statistics	Sawmilling
Research methods	Wood seasoning and preservation
	Fire control
	Mechanics
	Road construction

Silviculture will be dealt with only to the extent that it affects the production and economy of the forest enterprise and the organization of operations. Thus the variety of silvicultural systems used now or in the past will not be described or studied in any detail. That has been done adequately for Europe by Troup (1952) although subsequent recent trends have not been written up. But considerable reference to the need to have a dynamic silviculture which is both flexible yet practical will be made as well as the management organization necessary in the practice of typical silvicultural methods.

To clarify the scope of forest management one can classify the main problems that face the manager in three main groups:

1. *Control of the growing stock, its composition and structure*
    (i) Site adaptation.
    (ii) Choice of species and provenance.
    (iii) Manipulation of stands by cutting.
    (iv) Harvesting the produce.
    (v) Regeneration.
    (vi) Protection.
2. *Sales and distribution of the products*
   Communications, logging and sales.
3. *Operational efficiency*
   Administration of the property, organization and control of personnel and work, records and finance.

The chief peculiarities inherent in the management of forests compared with other undertakings and the basic influences on working forests are vested in the growing stock. It is on the constitution, growth capacity and situation of the growing stock that purpose mainly depends and control of the growing stock is the primary means of achieving the purpose.

Except in afforestation projects in which site preparation and choice of species are the primary problems, the chief activity in forestry is the manipulation or handling of the growing stock by cutting to collect the produce and at the same time to mould the forest stands into the forms, volumes and ages that are the most dynamic and suitable for the purpose of the owner. Even change of species or provenance requires first the felling of trees as does regeneration with the existing species. Aspects of protecting the growing stock also are frequently connected with the choice of when, where and what trees should be cut—for example in arranging shelter against wind or sun.

Manipulation of the growing stock by cutting requires constant decisions on four closely interlinked problems:

(i) How much to cut, to attain the desirable proportions of ages and sizes and relate the amount cut to the growth or increment of the forest.

(ii) When to cut, to favour the growth of trees left and also to collect the final harvest.

(iii) Where to cut, to aid decisions on (i) and (ii) and also to facilitate logging, needs of protection and the subsequent design of the forest structure.

(iv) How to cut, to achieve the intensity of fellings needed to promote the desired rapidity of growth of the trees left or to induce or encourage regeneration.

That control and moulding of the growing stock into the most desirable form, achieved by cutting, so that it contains trees having ages and sizes in the proportions needed to give the desired produce in regular quantities at the best advantageous rate in the prevailing conditions, is basic to all forest management. Discussion of these foundations of forest management will, therefore, form a substantial part of this book.

It is obvious that decisions on how to treat a forest and its growing stock cannot be made without an intimate knowledge of what it consists and of its potential capacity for growth. The manager must know what volumes of timber exist and how those volumes are distributed among stands of different ages or sizes of tree; for example he must know what proportion of the forest is mature. Additionally he must know the rates at which the trees are growing and will grow after treatment. But that is not all; he will also want to know whether other species could give better results in the sites available and whether and how those sites can be made more fertile. Knowing all those things he can judge what desirable form the growing stock should have to satisfy the objects of the owner and then plan how to handle the forest to lead it to the desirable state.

The planning therefore depends on a full assessment of sites and growing stock and their potential, interpreted in the light of other assessments such as markets, finance and labour resources or other constraints. But knowledge is never complete and circumstances change so that planning may have unforeseen results and can never be static. Repeated or even continuous assessments are, therefore, essential not only to collect new knowledge but to link results with treatment so that experience may improve planning and action.

The forest manager, therefore, is dependent on aid from the sciences of mensuration of trees and woods, of ecology, pedology, and economics for appreciation of repeated surveys and stocktakings. Those studies must be co-ordinated with records of past and current out-turn so that results of treatment can be assessed and the success of treatment judged. Organization of mensurational and other surveys and the incorporation of a system of records integrated with work as it is done are important elements of forest management.

This emphasis on management of the growing stock is not intended to belittle the importance of administration, logging, sales and communications. They too greatly affect forest management and in some circumstances may override management of the growing stock; for example, to raise capital and improve communications may be the first essentials in the early development of an inaccessible, remote forest. They also affect the daily management of the forest and may occupy much of the foresters' time and together control the efficiency of work. In this book, however, they will be discussed only to the extent that they affect policy, the management of the growing stock and the organization of the property.

Bibliography

Gordon, W. A. (1965) 'Sustained Yield and the Law', *Quarterly Journal of Forestry*, LIX (1)

Troup, R. S. (1952) *Silvicultural Systems*, 2nd edition, edited by E. W. Jones. The Clarendon Press, Oxford

Chapter II

OBJECTS OF MANAGEMENT

PURPOSE AND POLICY

Before considering the objects of forest management it is necessary to consider and establish clearly the difference between the basic purpose of an enterprise and the policy or course of action adopted to fulfil the purpose.

Objects of management are those which express a broad, basic purpose rather than a specific product to be produced. Examples of purpose are:

Attainment of the greatest profit.
Supply of goods for the benefit of a parent or essential industry.
Supply of services to the community.
Satisfaction of local needs for goods or employment.

Having decided the prime purpose of management, the policy of fulfilling the purpose can be settled. That policy will include definition of the kind of product to be made and the main methods of production. For example, to attain its purpose of attaining maximum profit a firm may decide that its resources in skills and capital and the prospects of demand make the catering trade attractive. Then, having decided to adopt the catering trade, it must choose its main methods—for example the wholesale manufacture of a product such as biscuits or the retail distribution of catering products bought from manufacturers. So, having the purpose of attaining profit the firm may decide on a long-term policy of wholesale manufacture of biscuits in a large factory centred in a highly populated region. Let us note that the biscuits are not an object of management but the means of achieving the purpose of highest profit. The policy of biscuit-making may change with circumstance but the purpose of attaining maximum profit will remain unchanged. But it is perhaps legitimate to call the biscuit-making a specific, local object of management in an individual factory within the main object of highest profit for the firm as a whole (see p. 43 in this chapter).

It is obvious that there must be no confusion of purpose if an enterprise has several objects. It then becomes essential either to segregate those regions or units in which individual objects have complete priority or arrange the objects in an order of priority, according to local circumstances as necessary, so that the minor objects act as constraints on the major purpose. Let us take a simple example. The owner of a stretch of river may wish to enjoy the fishing himself and also obtain the maximum profit from letting the fishing to others. Which of those two conflicting purposes is to have priority? Or, where on the river does one object have priority and constrain or completely exclude the other?

It may be noticed that sustained yields have not been included in the examples of purpose. Sustained yields are usually a means to an end. Thus an owner may wish to obtain maximum profit and at the same time wish to provide regular local employment; a sustained yield will promote the latter purpose but may interfere with the former at least temporarily; it is a means of satisfying the latter purpose not a purpose in itself, and a constraint on both purposes. Again the purpose may be to supply an industry or community with the forest goods that it requires; the *consumer* may need steady supplies so that sustained yields become a necessary part of the methods of growing and supplying the goods. Similarly the practical advantages of sustained yields in organizing work programmes, in use of equipment and man-power may be such that they cheapen management and promote sales so that efficiency and profits increase—the sustained yield becomes a means to the end of achieving eventual maximum profit.

On occasion, however, sustained yields may constitute a true purpose; thus an owner may desire maximum profit provided that his income does not unduly fluctuate. He may be compelled to accept lower profits to attain the essential purpose of a regular income so that the purpose of sustained yields becomes a constraint on the main purpose of maximum profit.

The policy or course of action chosen for an enterprise can be classed as either long-term or short-term. The former, as its title suggests, must be soundly conceived so that change is unlikely for many years. It will therefore be concerned with the main type of produce to be made, e.g. biscuits or furniture, which will control the whole type of manufacturing plant, and the positioning of the business. Such matters can only be changed with difficulty and at a high cost. The short-term policy or plan is the fulfilment of the long-term policy during the next few years—the immediate plan of action.

So there are three main phases in planning:

(i) The primary objects or purpose of the enterprise, segregated regionally and arranged in an order of priority, variable as necessary and according to local conditions;

(ii) The long-term policy for fulfilment of the primary objects;

(iii) The short-term plan for executing the policy in the immediate future.

Choice of objects and determination of policy

Choice of the objects of management may be comparatively easy for an industrial firm which is unlikely to consider objects other than profit except as subsidiary constraints. But for properties or concerns owned by the State, particularly for properties such as land which can provide a wide range of products and services, the choice may be hard to make. To what extent, for example, should land be devoted to the supply of services or goods locally consumed? Should the goods or services be provided at a profit or at cost, even free, at the expense of the general taxpayer to the benefit of the local neighbourhood?

Decisions on policy, the course of action to be taken, may be even harder to make, whoever may be the owner of the enterprise. Moreover policy will change as conditions change to a much greater extent than will primary purpose, so policy must be kept under constant revision in harmony with both conditions and objects which themselves may change but in degree and priority or locally rather than in substance.

Difficulties in deciding both the purpose and policy of an enterprise will naturally increase with the size and complexity of the project, with the degree of variety of its potential products and their importance to consumers, with the extent of the distribution of its activities and with the complexity of its owning and governing body. To that list must be added other concomitant factors such as trends in population changes, the march of invention and discovery of resources and consequent changes in standards of living all of which affect the pattern of demand, and even the chances of accident and catastrophe, local or world-wide.

The aims of forestry

In forestry it may not be too difficult to settle the main purpose or policy in broad terms which are themselves expressed in order of their importance and individual application to different localities. But definition of policy, e.g. specific material products to be grown, may be difficult owing to the long production period which necessitates shrewd forecasts of the trends of demand and which delays the effect of making any change needed to provide different products.

The fundamental purpose of forestry may be defined as that which secures the greatest continued value from the land allotted to forestry. In most circumstances that purpose might be extended by inserting the words 'widest and' before 'greatest' so that the largest possible number of people are included in the benefits obtained. Several authors have expressed this fundamental purpose in a similar way, such as: 'The objects of management under any circumstances is the most advantageous utilization possible of the soil allotted to forestry' (Knuchel, 1953) or 'The primary object of good management is the provision of the maximum benefit to the greatest number of people for all time' (Brasnett, 1953).

The values of forests

If we accept the definition now suggested it is necessary first to consider briefly what values, in products or services, forestry can offer from its use of land, for which it has to compete with other uses such as habitation, agriculture and other industry. Then the attitudes of different owning bodies to the various values can be assessed.

TYPES OF FOREST PRODUCE

Forest products may be placed in three main categories within two basic classes of produce—the material or tangible goods and the intangible services, as shown below:

A. *Intangible service products*
   1. Protective and regulative services.
   2. Socio-cultural services.
B. *Tangible material products*
   3. Material goods.

The protective and regulative services

The protective and regulative faculties of trees, particularly of diverse populations of trees such as forests, woods or copses, together with their company of ground vegetation, depend on three main features possessed by trees and vegetation which together create the environment of a forest:

(i) Their physical form and bulk including the network of their roots.

(ii) Their mode of living by absorbing water from the soil and transpiring it into the air.

(iii) The character of the decaying litter from leaves and branches shed on the surface soil with which it mingles.

It is evident that the very form and bulk of trees and their intertwining roots must check the movements of soil, air and snow and surface water, so that woods and forests are admirable controllers of land or snow-slips and of soil erosion, whether caused by movements of wind or water, and in many circumstances may even prevent them from occurring.

Clearly, undisturbed forests that have continuous and unbroken canopy perform those protective functions better than those interrupted by severe fellings.

The barrier formed by trees and their understorey also reduces the velocity of wind mainly to leeward but also for a shorter distance to windward. A succession of belts or scattered clumps and copses of trees can afford great protection from high, cold or desiccatory winds and also provide shade, thus providing a service which is invaluble for agriculture and animal husbandry in many climates.

The characteristics of trees and vegetation also affect the influence of rain and other precipitation. This is not the place to examine the many complications involved. But, in brief summary, it may be said that the water balance between the atmosphere and ground is expressed by the formula: Precipitation = Evapo-transpiration+Run-off+Percolation.

By increasing evapo-transpiration vegetation reduces, perhaps materially, the flow of water to water courses and reservoirs compared with that from bare land. The absorption of water (from the throughfall) by decaying vegetative litter also reduces run-off but increases percolation which is also aided by root channels. In the shade of vegetation snow melts more slowly so that the melt-water is unlikely to exceed the absorptive capacity of the litter and soil.

It is clear, therefore, that although vegetation by its comparatively high rate of evapo-transpiration reduces supplies of water to water courses and reservoirs it also reduces or may even eliminate run-off and thereby reduce erosion and, by delay, regulate the water-flow to water-courses. Floods are therefore checked or even prevented.

The denser and more productive the vegetation, such as forest, the greater are the effects. Also the more extreme the conditions, e.g. in localities that experience violent rain storms or have a steeply accented terrain, the more useful is this regulating power.

The effects of forests on amounts of rainfall is probably small. The extra height of trees above the land has a negligible effect on orographic rain. Nor can the greater humidity and possible lower temperatures around a forest have no more than a minor effect on precipitation. Cyclonic rain is not affected by forests. But in certain circumstances forests may appreciably increase fog-drip, or occult precipitation, by the capture of water droplets on the twigs of vegetation. (For expansion of this subject see Kittridge (1948).)

Socio-cultural services

Forests provide many social benefits which grow in importance with urbanization, improved transport, higher standards of living and general pressure on the land. There are three main aspects—employment, recreation and scientific study.

Employment provided by forests is valuable not only in providing a living but also in encouraging people to remain in or move to underpopulated tracts, and thereby improve the distribution of population and wealth. The employment involved is not merely for those working in the forests themselves, but also for those in the industries and trades dependent on forests. The need for encouraging this spread of employment may be so great that afforestation for that purpose may be part of a national policy, as in Great Britain.

The demands for open-air recreation and contact with nature is increasing rapidly in conjunction with urbanization and development of easy transport. Whereas until recently recreation in forests or other open country was confined to the few, now increasing thousands or even millions flock to the countryside in all heavily populated countries which have a high standard of living. The type of recreation wanted varies from remote luxury hotels, with facilities for water and other sports, to the simple picnic; from hunting, shooting or fishing to the afternoon walker or camper in a tent. There is no doubt that recreation, with which is included enjoyment of a forest just for its beauty and quiet, is becoming the most important benefit that some forests can provide.

Forests also provide opportunities for biological study of natural fauna and flora, of soils and hydrology. The heavy and increasing pressure on land in industrially developed countries gravely decreases the number and area of sites suitable for such scientific study. The provision of opportunities for studies of this kind, beyond those made by foresters themselves, has to be included in the objects of management of parts of some forests.

The owner's attitude to intangible benefits

Before considering the tangible products of forestry it is useful to examine the owner's approach to the intangible products or services when deciding his objects of management. The approach is somewhat different from that to material goods.

Before deciding the service objects of management the owner should assess three main things for each service or allied group of services:

    (i) The importance and value of each service both to the owner and the community.

    (ii) The extent to which the service can be provided by forestry or by some other activity.

    (iii) The cost (or profit) of providing each service, by forestry or by some other use of the land.

The study should enable the owner or owning body to determine, firstly, whether forestry or some other land use is best

2

suited to supply the services in whole or in part. That is a fundamental decision of policy on land use, a decision which should be lasting to prevent discontinuity and change and consequent ineffective execution. Secondly, for areas allotted to forestry, the owner should be enabled to decide what priority is to be given to the various services, both among themselves and relative to material forest products. He should also be able to decide the degree to which the supply of the services should be segregated to particular areas or combined with the supply of other services or products in one area.

The essence of the matter and yet its greatest difficulty is the assessment of values and costs in comparable real terms, e.g. in money. The value may be in negative yet concrete terms; thus failure to provide tree shelter-belts against wind might decrease agricultural products by 10 per cent in spite of a 5 per cent gain in area of agricultural land from forest. The social services are yet harder to value unless payment for them can be collected or the cost of providing them can be calculated. But one hopes that the cost is less than the value given. Is the value of a cathedral or welfare centre merely its cost?

Costs of providing services can often be judged more easily by estimating the value of what the land could produce in material goods were it not used for ensuring the service than by calculating the actual cost of administering the service. The enjoyed beauty of a wood may depend on its particular variety which includes open glades, overmature old hardwoods, some old conifers and a minimum amount of dense productive and profitable conifers arranged in convenient workable blocks. The loss in income from timber products incurred by keeping the forest beautiful will then be heavy although the cost of maintenance may be small. So the cost is mainly in terms of the loss of land that could have been used for efficient timber production. On the other hand let us note that there is still much beauty in woods managed entirely for timber production especially if large timber is grown.

An example of the cost of growing broadleaved species for the preservation of amenity in a forest where oak has been a traditional crop for centuries is given in the working plan (Anon. 1960), for the State forests of Dean and High Meadow in England where the preservation of amenity is a declared object of management as a constraint on the primary purpose of commercial profit. The cost as calculated by discounting to the present time at 3½ per cent compound interest the net revenues obtainable from growing oak, beech and various conifers. It was accordingly estimated that by foregoing the extra profits obtainable from growing conifers, to grow oak would cost from £2380 to £6545 a year over 1700 acres, or from £1·4 to £3·8 per acre per year, the actual amount depending on which of the alternative species of conifers were grown. To grow beech would cost less, namely from £595 to £4760 a year for the 1700 acres, or from £0·35 to £2·8 per acre per year.

The decision of the owner therefore requires not only a choice of objects but an allotment of priorities to the objects, priorities which may well vary from place to place in the same forest estate according to the suitability of each site for the various objects.

Material goods

There is a very large array of material forest products or goods which range from all the major products (i.e. all wooden products) to all the minor products (i.e. those which are not wood, such as resin, fruits or grass, whether they are more valuable than wooden products, such as firewood, or not).

Choice of what type of products to grow depends mainly on the kind of owner and his purpose and certain limitations and constraints. But two essential factors will always apply, namely:

(i) What products can be grown in the locality and at what continuous rate.
(ii) To what use the products can be put to the best advantage.

The first is an obvious limiting influence. Even if a valuable species and product can grow in a locality but only slowly, the delay and consequent sterilization of investment capital for a long time may make it disadvantageous to grow it. The rate of growth must also continue and not subside after perhaps one rotation.

The second factor is also clearly basic and may be more difficult to assess than the first because it will involve forecasting the trend of demand. But firstly the owner must decide what meaning he gives to the words 'best advantage'. The choice lies between three meanings, namely:

(i) Production of a particular kind of product for a particular consumer.
(ii) Attainment of the most desirable financial results irrespective of any particular type of product.
(iii) Production of the greatest quantity of products irrespective of the degree of financial gain.

Two examples of applying the first of those three meanings will clarify its scope. A community which owns forest might decide that it was essential to grow firewood because other forms of fuel would be unobtainable or too expensive to get and distribute in the forseeable future. Secondly, a paper-making company might decide that its interests were best served by growing only those species and sizes most suitable for its pulping plant; to do so might result in less profit from its forests but more profit from its paper project, its primary activity, so that the combined profit is enhanced. Other examples can easily be imagined and it is clear that the type of owner or owning body may substantially influence choice of this object of growing a particular product or products.

Different owners may ascribe different meanings to the most desirable financial results. What is the most desirable type of gain? Some owners may want the highest net, or even the highest gross income irrespective of the capital employed. Others may demand the highest financial yield, namely the highest rate of net income of the investment. Others again may be satisfied with a certain minimum rate of net income on the invested capital.

Factors that influence choice from the several basic objects of management will be considered below under ownership.

REGULATION OF YIELD

We have not yet considered the implication of the words 'continued value from the land' in the definition of the basic purpose of forestry proposed at the beginning of this chapter.

The attitude of forest owners to that concept will vary. It is obvious that if the primary aim is to supply a particular material for the sustenance of some industry—such as pulpwood for a paper mill—continuous regular annual, even daily, supplies may be essential. But by some owners, in some circumstances, variable supplies may be preferred, variable both in time and quantity, in the same way as variable amounts may be drawn from a banking account to suit variations in expenditure.

Moreover, to attain regular uniform annual yields from a forest will almost certainly require from time to time the cutting of some trees either before or after they are ripe for harvesting. To do that will interfere with the financial results so that there is a conflict between two probable objects of management, uniform yields and highest financial profit. Which of them is to have priority and to what extent?

These matters will be studied in some detail in the next and other chapters. But it may be accepted here that the insurance of continuous supplies of products in the sense of maintaining unfailing productive capacity, is always an object of forest management although uniform annual yields may not be necessary or even desirable. It is essential to maintain or improve the productive capacity of the machine—here the soil and healthy tree vegetation. To clear cut or cream a forest of its valuable trees and then leave the land untended or use it for some other purpose is not forestry but timber exploitation succeeded by neglect of the land or its appointment to some other land use.

Related to sustained yield there is another factor which may also be an important element in the purpose of management, namely whether a reserve of forest products should be accumulated to meet unforeseen emergencies. In other words, should less of the products be enjoyed now for future security? As will be explained in chapter III the reserve may take the form of either a reserve of living and growing forest products, e.g. timber trees, or a money reserve. Either type entails the use of a greater capital invested in the forest enterprise. To do so may reduce the rate of return on the capital. Does the owner wish to do that to ensure against future events?

In deciding the objects of management the owner cannot neglect consideration of what future demands for forest products may be. How will changes in population and living standards, development of invention and substitutes for forest products, affect the shape of demand on forests?

But one must be careful to distinguish between the purpose

and the means or methods used in achieving the purpose. Nevertheless, the means and methods necessary to achieve a purpose may affect the choice of that purpose owing to the difficulty and expense of the means and methods without adequate reward.

Basically the owner has to choose between three main things: services, a particular type of tangible product, or financial benefit (and the extent and interaction of each). If his choice of purpose is financial benefit then the trend of demand will affect his policy of the means and methods, i.e. what kind of product to grow—and not his main purpose unless the trend of demand seems so unfavourable that he decides to use his land for something other than forestry.

If, however, his choice is to produce a type of product to feed some other enterprise or class of consumer, his object depends on the future of that enterprise or needs of the consumer. Then his object will depend on future demands. Thus the present essential need for wood fuel in an inaccessible underdeveloped region may vanish if development were to introduce electric or other energy; the original purpose of supplying the people with their essential need for wood fuel will be changed.

In the same way increases of population, urbanization and ease of transport may involve a great future increase in the demand for forest services, the means to satisfy which must be considered and the object accepted beforehand. The decision is particularly important as it will affect other objects of management such as financial benefit, perhaps seriously.

THE INFLUENCE OF OWNERSHIP

In considering the effect of ownership on choice of purpose we may place owners in four main categories:

(i) Individuals.
(ii) Companies.
(iii) Communities or corporations.
(iv) Nations or States.

The desires of owners are differentiated and conditioned by breadth of interest, extent of resources, constraints or stimuli that apply and, dependent mainly on breadth of interest and resources available, the length of time that the owner can spare to achieve results. Each of those four elements will affect each of the four types of owner to a varying degree. But we can assume that every owner wishes to use the full potential of his land as far as his resources will allow.

Let us first consider the largest type of owner, the nation or state, not merely because it is the largest and most complicated but because its attitude to forestry and consequent policy must affect the other smaller owners within the state.

The land is an ultimate resource. It can seldom be extended—at least appreciably—and may even get less, by sea erosion. Therefore every state should have a land use policy within which a forest policy is fitted. Even if no land use policy is expressed a forest policy can and should be decided to guide and guard the selection and subsequent use of land for forestry. The earlier a forest (and land use) policy is formed in the development of a country the better it is; to change an adopted pattern of land use is difficult, expensive and a cause of hardship to many.

The interests of the State are very wide but its one purpose must be the prosperity of the whole people. Its attitude towards land, therefore, will be to get from it the greatest value for the greatest number of people. So it may be assumed that agricultural land use will be given priority over forestry wherever it is capable of the profitable production of food—except where and to the extent that forests are essential for aiding agriculture (by shelter belts for example), for giving protection against erosion or landslips or avalanches or for providing recreation facilities.

The social services of forestry, therefore, must always carry great weight, from services of providing employment in underpopulated tracts unsuitable for agriculture to protection and recreation. In some places such services may be more important than any other land use. There forestry will be given priority over other uses of land, whether the forests can be managed to supply tangible raw materials as well as the services or not. Elsewhere it will be the production of forest material goods and the commercial profit of the business that will influence use of the land.

Industry and other consumers want their raw material to be as cheap and good as possible. They buy in the most favourable market. Home grown forest goods must therefore be competitive with foreign goods in quality, price and convenience of supply. If the home product is not competitive forest raw material will be imported or, conversely, export markets will be lost. That would not only have an adverse effect on the balance of payments but also result in partial use of existing forest land and neglect of other land suitable only for forestry—in other words land would not be used to its capacity. But if that land could be used for some other more productive purpose and if imports of forest raw materials could be relied on continuously at reasonable prices and could be paid for, that would not matter.

But those suppositions are unlikely to hold. Rapidly rising populations ensure an increasing consumption of both wood products and food. The world area of forest is likely to fall progressively because all land which is sufficiently fertile and cultivable will be wanted for food production. Substitutes will meet some demands for wood but the properties of trees and its vegetable fibres are so variable and so useful that the demand for wood, either in its raw or processed form, will almost certainly increase markedly in the long forseeable future. There is no doubt therefore that a government should plan for an increase in the demand for forest products in the face of an increasing pressure on the land. That pressure, influenced greatly by short-term prospects, is likely to mean a shortage of forest land and products later—a shortage that the slow growth of trees makes it impossible to remedy in less than perhaps fifty years. Is a government prepared to face either a current demand for imports or reduced ability to export timber in return for current and perhaps temporary more profitable use of land and subsequent difficulty? Or alternatively expressed, are present lower profits worthwhile for subsequent good?

A government should therefore view the forests and land suitable for forest jealously. Moreover, wherever the balance of payments is critical the industries dependent on forests assume greater importance. The profit of the processing industry is added to the value of unprocessed exports and subtracted from the value of processed imports to the benefit of the home country—provided that the labour and capital to support the industry are available.

In certain circumstances a government may wish to retain a large reserve of growing timber for use in times of emergency.

That became the policy of the British Government in 1919 after the catastrophic effect of the 1914–18 war on timber and other imports. It was then decided to pursue a policy of afforestation and improved management of existing forests so that supplies of timber which would last for a three-year emergency could be accumulated. That policy continued until 1958 when it was considered that the implications of nuclear war made the prospects of a long emergency improbable.

To summarize, the main matters that influence governments in framing a forest policy are:

(i) The value of forest services—values which will increase in the future and may be paramount in some places.

(ii) The value of forest raw material owing to:

    (a) The necessity for wood in a wide range of use.

    (b) Increasing world demands for wood from smaller areas of forest.

    (c) The importance of the balance of payments.

(iii) The slow growth of trees to marketable size.

BRITISH FOREST POLICY

To illustrate a forest policy a summary of the main points of the statement made in 1963 on British Forest Policy by the Minister of Agriculture, Fisheries and Food has interest.

Through its Forest Service the British Government aimed at afforesting another 450,000 acres in the next decade. Afforestation would be concentrated in upland areas, where population is declining, to bring social and employment benefits. Afforestation in other areas would be where there are good economic reasons or where forests can maintain or improve the beauty of the landscape. In all forestry programmes more attention would be paid to provide public access and recreation and increase the beauty of the landscape.

Private forestry would continue to be supported by the existing arrangement of grants.

The steady expansion in both public and private forestry would mean a growing volume of home-produced timber to the benefit of the balance of payments and would enable the timber trade to have confidence in developing.

2*

The economy of forests

That policy, beyond the values of forestry to services, balance of payment and trade development, refers to economic use of forest land. What interpretation should be placed on the word 'economic' for state owned forest land?

It is difficult to avoid the conclusion that the meaning should be the best possible return from the capital invested, or at the least an approved minimum. It might, however, be argued that the object should be the highest net income or even highest gross income, irrespective of the capital invested. That argument would be supported by the view that, to quote Knuchel (1953), 'the public forest should supply all the assortments which the market demands, especially large sized timber, even if that can only be achieved with a large growing stock, and in lower interest on the growing stock' and again 'in public forests the first thing to look for, moreover, is not a higher *interest rate* on the forest capital but a constantly greater timber production on every acre'. That view is also supported if one admits that by such a return the state is providing the greatest good, to the greatest number—the greater the value that comes out of a forest the more are the people as a whole benefited. But to do that means in effect that the actual consumers are being subsidized. That too might be justified if the state were to have unlimited funds so that investment in other more profitable ways were not hindered. Similarly, the growing of larger sizes, i.e. having a larger capital, might be justified if industry depended on those sizes and could not get them elsewhere or only from undesirable imports. That last argument really means that the reduction of imports is more desirable than economic working at home.

In general, however, economic working of forests must surely mean the attainment of the highest or, at the least, a reasonable minimum net return on the capital used. Nevertheless, if a large growing stock already exists—i.e. a growing stock of large and therefore old trees—then maintenance of it with a high net income may be wise. The investment has already been made.

Another interpretation of the word 'economic' may also be considered. If the land available for forestry is limited and the need for quantity of home forest products is very great, then the aim may be to obtain the greatest volume of wood products per acre per annum. As will be seen later in chapter IV that would usually require a growing stock capital smaller than for the greatest net income and probably larger than for the best return on the capital employed.

When incorporating a national forest policy within its objects of management, therefore, a state forest service will have to consider carefully what its economic aims are—aims which may vary to some extent from region to region and from forest to forest within a region.

Communities and corporations

Having a narrower outlook and more slender resources than the state a forest-owning community or corporation will have somewhat different aims. To supply a particular product for the well-being of important elements of the community may be important—firewood for example, or special quality timbers for a local industry. Recreation and amenity in all or part of the forest may have high priority. A steady annual output of products and income is likely to be desirable as well as the accumulation of reserves for emergencies—for example, for repairs of storm or fire damage. Slender resources will make highest net returns to capital invested attractive, provided that other main objects are attained. A large growing stock will be attractive only for service products or supply of a particular class of material. Reserves for bad times might be accumulated by contributions to a reserve fund rather than by an excessive growing stock volume.

Companies

A company, unless its ownership of forests is unconnected with its main business and merely co-incident with the land it owns, will have special objects suited to its business. A paper manufacturing firm will want to grow wood suitable for paper pulp as cheaply and in as large quantities as possible even if the management of the forests alone would be more profitable by growing products for some other market. Very large companies might wish to integrate forest yields with several activities—e.g. with pulp mills, sawmilling and exports. Co-ordination of all activities could result in high financial yields from each. But maximum profit from the undertaking as a whole is the purpose of management, the forests being only a part.

Private individuals

The aim of a private owner will be narrower and usually more affected by shortages of resources than a community or company. He also will be interested in beauty and recreation as far as he can afford them. He will also want the highest net income with the smallest expenditure. If he already possesses a large growing stock giving a high income he will be happy to retain it—unless by realizing some of it and reducing the volume of the growing stock he can invest the proceeds elsewhere to better advantage and still retain a good forest income. If he has to build up an efficient growing stock he will want to have the highest net return from as small an investment as possible. He may also wish to have a reserve for emergencies. So he may be willing to sacrifice the advantages of a steady annual income by cutting less than what it is possible to cut in good times so that he can cut more in hard times. He may in other words, prefer a fluctuating yield and income provided that his forest capital is never seriously depleted and remains always fully productive.

Application of purpose

Having defined its aims the owning body must then plan their application to each region, forest and, indeed, each part of each forest. The plan will require an analysis of the aims in relation to the factors of each locality to determine both its productive capacity and the type and extent of the demands for its potential products. It should then be possible to decide to what extent each locality can contribute to the several purposes and the priority that should be given to them. This elucidation and application of purpose to a particular forest unit will constitute the objects of management for that unit.

The role of the forest will depend on four basic elements:

(1) The growth potential—a combination of factors of climate and soils.
(2) The protective potential—a combination of factors of topography, climate, soils and situation.
(3) The accessibility—a combination of factors of topography and position.
(4) Constraints—arising from legislation, claims of right-holders, previous commitments and the like.

The first of those elements determines what kind of trees and vegetation can be grown, at what rate of growth and to what sizes and quality. The second and third elements determine the types of demand and the extent to which they are likely to affect the forest. The fourth element will determine the extent to which the use of the forest is controlled or limited and directed to a particular type of produce or consumer. For example, it may be essential to supply firewood to right-holders. An analysis of those elements will enable the owner, influenced by the present state of the forest estate, his own resources and the trends of demand in general, to determine what role the forest, and each part of the forest, is best fitted to play within his forest policy.

As the satisfaction of one object may interfere with the realization of another it is necessary to assign priorities when two or more objects are to be fulfilled in one and the same site. Thus, both regular sustained yields and the best net income may be wanted; to attain the former stands may have to be felled and harvested before or after the most profitable time; but to avoid that sacrifice fluctuating yields may be perpetuated. So the attainment of one object has to be qualified in favour of the other.

To improve efficiency and facilitate management it may be desirable to segregate different parts of the forest for different main objects. For example, the more accessible areas may be set apart for amenity management or, perhaps, for the production of wood fuel for the nearby community and the more distant parts to the production of saw-timber. In each part management can then concentrate on the one subject.

Separation of objects for different sites leads one to appreciate the need for subsidiary segregation of different courses of action even for one main object. Let us consider a forest in which three quarters of it are set apart for the attainment of the highest financial yield in a region where there is a good market for a variety of forest goods. Study of market prices and of site qualities may show that on some sites it is most profitable to grow poplars, on other sites pines for pulp and on the remainder a variety of conifers for saw-timber. Each of the three main site types could be segregated for separate treatment. The policy or course of action of management to fulfil the object of highest financial yield in the respective sites would be to grow poplar, pinewood pulp or coniferous saw-timber—each to the most profitable quality, size or age.

The main purpose of management has now been narrowed and clarified to a definite product to be grown—a specific policy for the locality concerned. Each of those specific policies is the method adopted to achieve the highest financial yield but may be termed and recognized as a specific, local object of management.

Such policies or specific objects of management are eminently desirable. By translating the main purpose to a precise and specific policy or local object for definite site types management is facilitated and made more efficient. In reality such specific, local objects of management are the means or policies by which to attain the basic purpose. Nevertheless they can be truly accepted as objects, even essential objects, in the same way as the object of a machine shop may be to make as efficiently as possible cylinder blocks for a motor car factory whose primary aim is to make a profit from the manufacture of motor cars.

The time factor

All objects of management should be as long-term as possible. Change of purpose entails a change in the means of attaining it and that requires a change in the constitution of the forest. The slow growth of trees makes change in the constitution of a forest laborious and slow; there will be a delay in achieving the new purpose. If that purpose changes again in the meanwhile confusion begins.

Nevertheless changes in markets occur. Specific objects of management which define the types of product to be grown will then have to change also. An example is the steady disappearance of markets in north-western Europe for fuel-wood and small hardwood timber for agricultural use. Hardwood coppices have progressively declined in value and for years have been undergoing conversion to high forest, often coniferous, causing an interruption in yields and income for many years in addition to the expense of the conversion itself.

Clearly it is necessary for objects of management to be selected with great care after diagnosis of market trends. Particular attention has to be given to the main types of demand— for example, hardwood or coniferous timber. To convert a forest from one species to another is more drastic and expensive than to change the sizes grown of one and the same species. It is clearly easier to change a forest growing coniferous saw-timber to one that grows coniferous pulp than to one growing hardwood saw-timber. In the former conversion, only a reduction of the average marketable size (and age) is likely to be necessary. To reduce the average age of a forest will involve more felling (more than the increment) and higher yields (and income) for some years. Conversely, to change from the production of pulp to saw-timber sizes of the same species would require the average age in the forest to be increased—by cutting less for a time with consequent and perhaps painful loss of income.

Change in the quality or form desired in timber products may be impossible to achieve without starting all over again. Open-grown, branchy and short boled oak, grown in England for the object of naval timber a hundred and more years ago and formed by heavy thinnings in youth, cannot now be changed in form by any subsequent treatment. The oak must be replaced with crops grown and tended to the new standards wanted.

The trend of world and national demands for forest produce is, therefore, an essential element in determining both the national forest policy and the specific objects of management.

The Food and Agricultural Organization has recently made a special study and published its results (Anon. 1966). It is pertinent to refer to the basic features of the study.

(i) The population of the world, about 2990 million in 1960, is expected to increase to 3907 million in 1975 and to 5965 million in the year 2000. Of this increase of 2975 million in forty years, over 2500 million will be in the developing countries—Latin America, Africa, and Asia (excluding Japan), i.e. more than doubling their present figure of 2043 million.

(ii) The world consumption of woody raw material is expected to increase from 2131 million cubic metres in 1960–62 to 2689 million in 1975, i.e. by 25 per cent.

(iii) The estimated world consumption of fuelwood will increase by only 10 per cent—from 1088 to 1999 million cubic metres by 1975. But in Europe, North America and the U.S.S.R. the consumption will fall materially and increase in Latin America, Africa, and Asia-Pacific, where however the consumption *per caput* is falling.

(iv) The consumption of industrial wood is expected to rise by 43 per cent—from 1043 to 1490 million cubic metres. The increase will occur in all regions but least in Africa and the Pacific.

(v) The increase in consumption of industrial wood will be mainly in pulpwood (by 267 million cubic metres) and less in saw and veneer logs (183 million cubic metres).

(vi) Although the 1960–62 volume consumption of round-wood (for posts, fences, and other unprocessed use) was fairly high at 188 million cubic metres, its general world use will decline slightly by 3 million cubic metres in 1975. The decline is confined to the developing countries although the consumption *per caput* is declining everywhere.

These trends may be simplified to the following:

(i) the need for material forest goods will go on increasing steadily.

(ii) the greatest increase will be in wood for processing—for sawing and veneers but particularly for pulp and particle board. The degree of increase will vary directly with increase of wealth and industrial development of a country.

(iii) the need for wood in the raw (fuel and round wood) will increase only in developing countries and there the consumption *per caput* will fall as wealth and industry develop.

The consequent aims of forestry (excluding those for recreation and services) should be:

(i) To strive for steadily increasing supplies of wood. Means of achievement include:

(*a*) Increasing the useful yearly volume growth in each unit of forest area.

(*b*) Improvement of road and other communications to forests hitherto difficult of access.

(*c*) Afforesting land available.

(ii) To match the production of wood with the various types of wood needed. That would entail mainly:

(*a*) A steadily increasing supply of timber suitable for paper and other wood-fibre materials. Quantity rather than quality of tree form would be the main aim.

(*b*) An increasing supply of saw and veneer timber. Quality of tree form rather than quantity would be the aim.

To widen and support that brief summary of the trend of the world's demand for wood, four tabular statements from the F.A.O. report are given below in Tables 1 to 4, which are reproduced from *Unasylva*, vol. 20 (1–2), nos. 80–81, 1966, with kind permission of the Food and Agriculture Organization, Rome.

Table 1

CHANGE IN RECORDED WORLD USE OF WOOD AND WOOD PRODUCTS, 1950–52 TO 1963

	Million units	1950–52	1955–57	1960–62	1963	Change 1951–61 Volume	Index
ROUNDWOOD							1951 — 100
Sawlogs[1] and veneer logs .	Cubic meters	493·4	582·2	648·3	656·7	+154·9	131
Pulpwood[2] and pitprops . .	Cubic meters	185·6	233·5	255·6	257·1	+ 70·0	138
Other industrial wood . .	Cubic meters	129·2	124·9	116·3	118·3	− 12·9	90
[4]*Total industrial wood .*	Cubic meters	808·2	940·6	1 020·2	1 032·1	+212·0	126
Fuelwood . .	Cubic meters	865·6	876·1	876·5	886·5	+ 10·9	101
TOTAL . .	Cubic meters	1 673·8	1 816·7	1 896·7	1 918·6	+222·9	113
WOOD PRODUCTS							
Sawnwood[3] .	Cubic meters	266·1	309·6	341·0	354·2	+ 74·9	128
Paper and paperboard . .	Metric tons	44·3	59·4	77·3	84·0	+ 33·0	174
Plywood . .	Cubic meters	6·8	11·3	16·8	20·1	+ 10·0	247
Fibreboard .	Metric tons	2·2	3·3	4·5	5·3	+ 2·3	210
Particle board .	Metric tons	0·04	0·57	2·29	3·56	+ 2·25	5 900
Roundwood[5] .	Cubic meters	[4]129·2	124·6	116·3	118·3	13·9	90

[1] Includes logs for sleepers. — [2] Includes roundwood used for the manufacture of particle board and fibreboard. — [4] Production. — [3] Includes sleepers. — [5] Excludes pitprops.

Table 2

THE STRUCTURE OF ESTIMATED TOTAL WOOD USE IN 1960–62

	Sawnwood (m3)	Panel products[1] (m3)	Pulp products[2] (MT)	Roundwood [m3 (r)]	Total industrial[3],[4] wood [m3 (r)]	Fuelwood [m3 (r)]
	. . . . Million units					
Europe	78·31	8·41	22·87	36·60	246·70	107·90
U.S.S.R.	99·72	2·24	3·47	67·30	249·70	100·90
North America . .	94·37	16·25	37·37	18·60	308·70	46·10
Latin America . .	12·39	0·52	2·66	8·50	38·70	192·40
Africa	4·05	0·37	0·90	13·30	23·50	182·70
Asia-Pacific . . .	57·33	2·75	10·20	44·10	176·00	457·60
WORLD TOTAL .	346·20	30·50	77·50	188·00	1 043·00	1 088·00

[1] Excludes veneer. — [2] Includes some nonwood fiber products, excludes dissolving pulp. — [3] The product quantities have been converted into equivalent volumes of roundwood using the standard factors published in FAO, *Yearbook of forest products statistics*. No allowance has been made at this stage for the variation in transformation ratio from region to region, for variation within an industry or over time; nor has the volume of wood raw material supplied in the form of wood residues been subtracted. These matters will be considered in Chapter IV. The present estimates of roundwood equivalent are intended to do no more than give a preliminary indication of the broad orders of magnitude of the wood raw material corresponding to the estimates of consumption of wood products presented in this chapter. — [3] Excludes wood raw material equivalents of veneer and dissolving pulp consumption.

Table 3

ESTIMATED CHANGE IN WORLD CONSUMPTION OF WOOD AND WOOD PRODUCTS, 1960–62 TO 1975

	Million units	1960–62	1975	Increase 1961–75 Volume	Index
WOOD PRODUCTS					1961 — 100
Sawnwood	Cubic meters (s)	346·20	427·30	81·10	123
Pulp products[1]	Metric tons	77·50	161·90	84·40	209
Panel products[2] . . .	Cubic meters (s)	30·50	75·80	45·30	248
Roundwood	Cubic meters (r)	188·00	185·00	−3·00	98
Fuelwood	Cubic meters (r)	1 088·00	1 199·00	111·00	110
WOOD RAW MATERIAL[3]					
Sawlogs and veneer logs .	Cubic meters (r)	629·00	815·00	186·00	129
Pulpwood[4]	Cubic meters (r)	226·00	493·00	267·00	218
Other industrial wood . .	Cubic meters (r)	188·00	185·00	−3·00	98
Total industrial wood .	Cubic meters (r)	1 043·00	1 493·00	450·00	143
Fuelwood	Cubic meters (r)	1 088·00	1 199·00	111·00	110
TOTAL	Cubic metres (r)	2 131·00	2 692·00	561·00	126

[1] Excludes dissolving pulp, including some nonwood fiber products. — [2] Excludes veneer. — [3] See footnote 3 in Table II-2. — [4] Includes roundwood used for the manufacture of particle board and fiberboard.

Table 4

ESTIMATED CHANGE IN ANNUAL WOOD CONSUMPTION BETWEEN 1960–62 AND 1975

	Sawn-wood	Panel[2] products	Pulp[3] products	Round-wood	Total industrial wood	Fuelwood	Total
	. . . . Million cubic meters wood raw material equivalents[1]						
EUROPE	+15·1	+20·9	+76·1	− 12·8	+ 99·3	−33·9	+ 65·4
of which:							
Northwestern Europe[4]	+ 9·7	+12·1	+49·2	− 5·3	+ 65·7	− 12·4	53·3
U.S.S.R.	+19·8	+19·0	+31·6	− 7·4	+ 63·0	−20·9	+ 42·1
NORTH AMERICA . .	+23·4	+22·0	+51·4	− 7·6	+ 89·2	−12·1	+ 77·1
of which:							
United States . .	+21·2	+19·7	+46·7	− 7·3	+ 80·3	−10·7	+ 69·6
LATIN AMERICA . .	+21·7	+ 2·1	+11·6	+ 3·6	+ 39·0	+27·6	+ 66·6
AFRICA	+ 6·1	+ 1·6	+ 4·5	+ 3·2	+ 15·4	+63·3	+ 78·7
of which:							
Southern Africa .	+ 0·3	. .	+ 1·5	− 0·6	+ 1·2	+ 0·5	+ 1·7
ASIA-PACIFIC . . .	+58·6	+11·9	+55·8	+17·6	+143·9	+86·9	+230·8
of which:							
Japan	+20·8	+ 6·1	+26·1	+ 0·9	+ 53·9	− 2·7	+ 51·2
China (Mainland) .	+16·4	+ 2·7	+16·1	+ 5·5	+ 40·7	+10·0	+ 51·7
Pacific	+ 3·0	+ 0·2	+ 3·0	+ 0·0	+ 6·2	− 1·4	+ 4·8

[1] See footnote 3 in Table 2. [2] Excluding veneer. — [3] Exceeding dissolving pulp. — [4] EEC, United Kingdom and Ireland.

Bibliography

Anon (1960) 'Working Plan for Dean and High Meadow.' Forestry Commission of Great Britain (internal circulation)

Anon (1966) 'Wood; World Trends and Prospects', *Unasylva*, 20 (1–2), nos 80–81

Brasnett, N. V. (1953) *Planned Management of Forests.* George Allen and Unwin, London

Kittridge, J. (1948) *Forest Influences.* McGraw-Hill Book Co., Inc., New York

Knuchel, H. (1953) *Planning and Control in the Managed Forest*, translated by M. L. Anderson. Oliver and Boyd, Edinburgh

Chapter III

SUSTAINED YIELDS

It is fair to say that the practice of forestry began with the silvicultural systems of simple coppice and coppice with standards. Previously forests had certainly been cleared for village sites and agriculture and the timber used or burnt; there had also been multiple use of accessible forest—trees and vegetation were cut and removed for some definite use, fruits and edible roots collected and consumed, animals hunted for food. But in early times those uses of forests constituted no conscious management of them, but were merely exploitation of forest resources which were apparently inexhaustible.

But as a pattern of villages and towns surrounded by expanding agriculture evolved, the ability of most hardwood species to coppice would have been appreciated. When the new growth reached usable size it would have been cut again and the process repeated. As the size of material wanted would have been small—for firewood, poles and posts—the cycle of events occurred in the memory of a man. A relation would soon have been established between the time taken to grow requirements and the area of forest needed to supply them. Also as stools lost vigour with old age and as regeneration got damaged by grazing cattle the need to ensure both regeneration from seed (by leaving standards or reserves) and its protection was realized. Examples are the Forest Acts of 1482 and 1543 in England, and the Ordonnance de Mélun of 1376 in France (for greater detail see Chapter X). The need for regular continuous yields, attainable by protection, regeneration and maintenance of adequate areas of forest was slowly realized.

But in Europe it was not until the middle of the eighteenth and beginning of the nineteenth century that principles of forest management began to be fully expounded and established to succeed the silvicultural and managerial success of the Tire et Aire methods that had been practised in France in hardwood high forest, since the sixteenth century. German foresters

(Öttelt, Georg Hartig and Heinrich von Cotta) then propounded the organization of silviculture to provide sustained yields of the produce desired in the most practical way and for financial advantage (see chapters IV, VIII and X).

Before studying the effect of a sustained yield on the organization and structure of forests it is necessary to examine the definition of a sustained yield, the types of yield involved and the value of sustention.

The meaning of a sustained yield

A sustained yield may be defined as the regular, continuous supply of the desired produce to the full capacity of the forest.

That definition clearly includes all forest products, the intangible and the tangible, protection and amenity, major products (wood) and minor products. The principles of sustained yield apply to all the products chosen for supply and management will be planned to integrate the supply of all to the best advantage of each. The arrangements may be to segregate the supply of each product from separate areas of forest; each individual area can then be worked for the sustained yield of the chosen product to the capacity of the site. But if several products are to be supplied from one area the production and yields of some products may interfere with those of other products. Yet sustained yields of all should be possible although in quantity and value the individual yields may be less than the potential yield of each separately.

The value of sustention for intangible forest products is no less and perhaps even greater than for the tangible. It may also be as great for desired minor products such as edible fruits as for major products such as firewood. But for clarity this chapter will deal mainly with major products and money yields but, *mutatis mutandis*, the yield of other produce also applies. The multiple use of forests must not be forgotten.

Types of sustained yields

Sustained yields of wood can be collected in three different ways which may be described as:

1. Integral yields when the whole forest contains only one age of tree so that the whole forest is felled and harvested (and then regenerated) at one time, e.g. once every fifty years.

2. Intermittent yields when there are several age classes of tree so that timber becomes ripe for felling at regular intervals of several years.

3. Annual yields when timber ready for felling is available every year.

Commonly, but rather loosely, sustained yields are often understood to comprise only annual sustention.

Factors affecting sustained yields

In all types of sustained yield the basis of its attainment is sustained production, i.e. sustained growth of the crop, which in turn depends on maintenance of the fertility of the site. No yield, whether it is collected each year, at regular or irregular intervals, can be continued indefinitely unless the production or growth of timber (or other produce) is maintained by regeneration and continued fertility of the soil. If growth of the desired produce is fully maintained even irregular periodic yields might be considered to provide a sustained but intermittent yield. At least production of the desired produce is sustained.

Moreover man is never satisfied with what he has but constantly wants more or better things. He will want steadily increasing yields from forest lands. Therefore the essence of the matter is both the sustention and improvement of the crops and the fertility of the land. Then the forest crop must be so tended that the desired produce is grown and regenerated with as little interference as possible from competing vegetation, disease and from physical damage from the weather, fire, animals or men. The regularity of the yield can then be influenced by suitable manipulation which ensures that a sequence of trees that have the desired sizes or ages are continuously available for felling in regular quantities to suit the objects of management.

Means of regulating sustained yields

1. *By area.* A simple example of a sustained yield is a wood of 10 acres uniformly and fully stocked with sweet chestnut worked as simple coppice which provides the desired produce at an age of ten years. One acre is cut in the winter of each year and automatically regenerated by coppice shoots in the succeeding spring. Provided that the site is uniformly fertile over the whole wood and provided the fertility is maintained, equal,

sustained, annual volume yields will be obtained. Expressed as an area yield the annual yield, $AY = A/R$ acres, where A is the area of the wood in acres and R is the rotation or interval between fellings.

2. *Reduced or equiproductive areas.* In a large wood it is unlikely that soils and sites will be uniform and the consequent variations of fertility will cause variations in growth rates. Then, although the areas cut each year will be the same, the volumes of wood yielded will differ. These variations in yearly volume yields can be mitigated by the device of using equiproductive or 'reduced' areas.

To use 'reduced' or equiproductive areas sites are classified according to their productive capacity. Thus if the most fertile sites, quality I, produce 100 cubic feet of timber per acre annually, quality II sites produce 80 cubic feet and quality III 50 cubic feet then to produce 100 cubic feet annually either one acre of Q I, or $1\frac{1}{4}$ acres of Q II or 2 acres of Q III are needed. So the actual acreages of Q I, II and III sites are divided by 1, 5/4 and 2 respectively to give the number of 'reduced' acres in each. Each reduced acre is equiproductive. Then $AY = Ae/R$, where Ae is the number of reduced acres.

3. *Other means.* Units other than area are often used to regulate annual yields, namely volume of timber or numbers and sizes of trees. The methods employed will be considered in detail in chapter VIII. But we can note now that it is usually unnecessary and even undesirable to equate annual yields; instead yields are regulated over a period of years. For example a rotation, or interval between regeneration and final felling, may be divided into several periods and equal areas of forest allotted to each period. Thus a rotation of sixty years might be divided into six periods; one-sixth of the area of the forest is thus felled consecutively in every ten-year period, although in any one year somewhat more or less than one-sixth may be cut as circumstances may indicate.

Justification of sustained yields

There can be no valid objection to sustained production, namely the sustention of forest land and forest crops in their continued most healthy and dynamic state. But objections to sustained yields, the amount of produce cut, and, more particularly, to yields rigidly regulated to prevent annual or periodic rises and falls in a period of years, can be and are raised.

Such objections have arisen comparatively recently and show a considerable change from earlier thinking which gave the object of sustained yields, strictly applied, an essential place in all proper forestry theory and practice. To understand this change or, rather, relaxation of opinion it is necessary to appreciate the change in conditions that apply in many parts of the world and particularly in the industrially developed countries.

At the end of the eighteenth century and beginning of the nineteenth when the necessity of sustained yields was propounded in Europe and universally accepted by foresters, communications and transport were bad. Villages and even towns were largely self-contained entities unless placed on a navigable water-way. They had to depend on locally produced goods except luxuries and a few expensive necessities that could not be made nearby. Power from electricity and oil for heat and light were unknown; coal was available only in the less remote country districts. Wood, bulky and cumbersome to transport, was a necessity nearly everywhere for fuel and agricultural uses and very desirable for furniture and house or other construction for which small, local sawmills provided sawn timber. Interruption of assured regular supplies of wood from a nearby source was a serious matter which upset the economy of the community.

Regular employment, ensured by sustained yields, was also an important matter which affected not only the efficiency of forest management but the welfare of the largely agricultural community. Spasmodic, irregular yields cause an ebb and flow in forest work; changes in the quantities felled affect all other work—from regeneration to extraction, from road repairs to silvicultural tending. Men had to walk to the job so that they lived within some two or three miles from it. Fluctuating needs for labour not only affected the community welfare but periods of unemployment encouraged migration of labour which would be difficult to replace when the need recurred.

Industrial markets were then individually small and local—sawmilling, cooperage, turnery and the like. All required regular supplies of wood from local forests; temporary arrangements for supplies from distant forests would be difficult or impossible to make.

Nor did there seem to be any undue difficulty in oganizing sustained yields. The types of wood wanted had remained unchanged for centuries—smallwood for fuel and agriculture, constructional sawn timber and big hardwood timber for shipbuilding. The pattern of demand seemed sure although the amounts wanted were rising with population. The increasing demand for pit-props and then for coniferous pulp-wood did not seem to affect the pattern seriously until, perhaps 1861 when the battle of Hampton Roads doomed big wooden ships.

Therefore there seemed to be no obstacle to, but only advantages in organizing rigid forestry plans, with a sustained yield for each forest as the primary purpose, involving definite rotations and even-aged stands. The aim was pursued even further so that parts of a forest were worked as separate management units so that each had its own sustained yield. Nor did there seem any objection to sustained yields from uneven-age forests except that the means to achieve them were not then understood. It was, however, recognized that the attainment of more or less equal annual sustained yields might take a long time and would often require some sacrifice in felling stands before or after they were mature. But the final goal of sustained yields from each forest and its separate parts was deemed to be paramount.

In modern times changes in two main factors, transport and markets, have changed thinking. There is no doubt that of the many outstanding developments in transport it is the perfection of the internal combustion engine and consequent elaboration and improvement of roads that have had the greatest effect on forestry. A network of roads and tracks in developed forests enables trucks to penetrate the felling sites where they can be loaded by other mechanical equipment. The loads can then be taken economically and quickly, within the day, to markets 50 or more miles away. No longer, therefore, need markets depend on supplies from a local forest but from forests within a radius of 50 miles.

Bicycles and motor cycles or cars now make it easier for men to go 10 to 15 miles to work than it was to walk 2 or 3 miles sixty or seventy years ago. Moreover, a large owner such as the

State can take the labour daily to and from the work in lorries up to some 20 miles. Surplus labour in one forest or part of a forest where yields and work are temporarily low can be taken daily to another site where work may be concentrated. In an industrially developed country there is no essential need, therefore, for separate parts of a forest to have sustained yields either for convenience of those employed or for local markets.

Two features of modern markets also affect thinking on sustained yields. The first is the greater concentration of industry which uses forest products. The great wood-fibre using industry for paper pulp and panel products requires large plant, a big output and a heavy consumption of forest produce at one place. Sawmills also have grown larger, perhaps integrated with pulp mills. Other small users of timber, such as turnery and co-operage works, have either disappeared or become larger. Those changes, allied with easy transport, make forest industry dependent on supplies of raw material from a region, from a consortium of forests or from a very large forest rather than from small individual forests or parts of a forest. Supplies to industry must still be sustained but from a wider source than previously and may even include imports from another country, imports which can either provide a reserve against temporary local shortages of supplies or be, as in Britain, the main source of raw material which is supplemented locally.

Changes in recent years in the pattern of demand, such as that from small hardwood posts and poles used in the raw to coniferous pulp-wood, have made foresters less certain of the future. Now European hardwood coppice forests are being converted to high forest, often coniferous. Will future technological developments, exemplified by the present explosive increase in the manufacture of particle board, again require new radical changes, perhaps even a return to coppice? In consequence foresters must plan flexibly. In itself that does not reduce the importance of sustained yields from individual forests. But it does necessitate management for sustained, optimum growth and wise forecasts of demand and skilful, flexible management which will place less reliance on organizing yields calculated in advance for a planned period of years, for a definite rotation. There will be more frequent adjustments phased to prevent undesirable fluctuations in yield or to improve growth but not those designed to satisfy the primary purpose of attaining rigidly sustained annual yields.

Economic factors. There are also two economic objections to sustained yields. Firstly, rigidly regulated annual yields prevent an increase of felling and sales during time of high prices or a reduction for low prices. There is no modification of fellings to suit demand; consequently not only does the owner suffer but high prices tend to rise still higher and low prices to fall yet more. Such rigidity in felling programmes is obviously inadvisable but can be overcome by regulating yields over a period of years instead of yearly.

But a second economic objection is more serious. To provide sustained yields without unreasonable fluctuations from time to time, the forest must contain trees of different ages and sizes in the proportions necessary to produce a regular sequence of maturing timber. To mould the forest into that state will require the cutting of stands either before or after the financially most advantageous time. It may, for example, be necessary to husband the stock of slow-growing mature and over mature stands by reduced fellings until immature stands grow to maturity. Financially it would be better to realize the mature stands quickly by heavy fellings, replace them with vigorous young regeneration and endure a gap in yields until the younger stands reach maturity. But the heavy fellings made now and the proportionately heavy regeneration of them will perpetuate the mal-distribution of ages and sizes of trees in the forest. Yields will again fluctuate.

Reserves

Fluctuating yields of an individual forest may have financial advantages. But some owners, particularly private individuals or communities that own and manage forests for their particular needs, may depend on a regular annual forest income or supply of wood. To them sustained yields from a forest, regulated to uniformity over short periods, may be more advantageous than financial improvements that involve big fluctuating yields. But they are also liable to experience emergencies when increased forest yields can pay for the emergency or provide the material needed. In effect, therefore, they want both a sustained yield and a reserve on which they can draw when necessary. Without reserves there is the danger that the forest will be heavily exploited for the emergency without a corresponding reduction afterwards when the usual full demands on the forest continue. Repetition of such action would in time destroy the forest by over-cutting.

Timber reserves. Timber reserves can be formed in two ways. In the one way certain stands in the forest are isolated from ordinary management, i.e. from sustained yield working, and are allowed to grow on and accumulate volume subject only to silvicultural thinning and tending. They would then be available for felling in an emergency and may be termed 'standing' reserves. Unfortunately the emergencies may not coincide with a suitable maturity of the reserves. They are not satisfactory.

The other way is to accumulate 'movable' timber reserves by cutting somewhat less than the increment or rate of growth. The whole forest, with its managed proportions of ages and sizes of trees, then accumulates a larger volume or capital which is equivalent to the use of a rotation longer than that otherwise needed. It is then easy to cut more during a crisis and then replace the excess cut by reverting to the usual conservative yield and allowing the capital again to accumulate.

Movable reserves can therefore be a valuable insurance against inconvenient inflexibility that may result from the execution of a sustained yield policy, but the insurance is at the expense of using a somewhat greater growing stock capital. But movable reserves will not mitigate a drop in income caused by slack demand and low prices. It may not be possible to sell more timber even at reduced prices which, indeed, would be liable to move even lower with increased supplies to a stagnant market.

Money reserves. The solution is to accumulate a money reserve instead of a timber reserve. The reserve fund could be fed with regular contributions and with income from unusual receipts or from capital sources. Examples are unexpected income from high prices, receipts from excess fellings such as those from windfalls or other accidental damage, sales of parts of the forest estate and perhaps an annual contribution from a definite percentage of the usual forest income. Interest from the fund itself could be retained in the fund.

The fund could be used not only for emergency demands of the owning body but also for the capital improvement of the forest estate or repair of unusual damage sustained by it. Examples are purchase of land for afforestation, improvement of communications, repair of bridges and roads damaged by flood or land slip, reforestation of woods destroyed by fire and the like are examples.

Summary

In summary it can be seen that sustained yields from individual forests or from separately worked units in one and the same forest have not the same necessity as they had only fifty years ago, at least in industrially developed countries. Easy transport of goods and people make local sustention less important.

Nevertheless, the value of sustained yields from individual forests, especially for small owners or communities dependent on local resources for goods and employment, should not be underestimated. Organization of current and future work is facilitated by sustained yields, constant employment is offered so that a permanent, skilled labour force can be established, markets can be fostered and their confidence gained, deployment of work among supervisory staff is promoted so that they are not periodically over- or under-worked, use of mechanical equipment and of forest roads is steady, enabling replacement and repair programmes to be more easily planned and executed and silvicultural tending can be more easily organized. Together those factors, all promoting efficiency of management, provide cogent, convincing reasons for sustained yields in each forest that is an administrative unit.

For State owned forests, however, the need for sustained yields from individual forests is somewhat less. The State is concerned with fostering the general welfare and development of industry and general wealth. Success will be influenced not on the output of produce from a single forest but from all forests within economic range of transport, perhaps within a 50 mile radius. If the State is the major owner of properly managed forests a regional sustained yield will suffice although the yields from individual forests may fluctuate according to peculiarities of the growing stock in each. Nevertheless, it is probably easier and in the end economically more satisfactory and efficient if each individual forest (or local group of woods worked as an administrative unit) has its own sustained yield thereby also ensuring the sustained yield of the whole region.

There may be a prolonged period of adjustment in which a State is shaping its forest estate by acquisition and afforestation of land, purchase of forests and re-organization of existing State forests. That period may inevitably include large fluctuations of yield from individual forests and yet include steadily rising yields from a region and consortium of forests. But the final most desirable goal must surely be a sustained yield from each forest or group of woods forming separate administrative units. Sustained yields from individual private forests will obviously help to ensure regional sustained yields.

Nothing that has been written lessens the need for forests to be managed to give sustained vegetative production, whether the material output or yield of the forest is sustained or not. Sustained vegetative production must always be the aim. To attain it the soil of each unit of land must be kept as healthy and fertile as possible; indeed it should be improved. And on each unit of land the tree crops must be kept as vigorous as possible to produce as rapidly as they can the material wanted. To do that may entail heavy fellings and replacement of over-mature or other slow-growing stands and thereby delay the attainment of sustained yields in a particular forest. The delay may not affect a region and it remains true that until vegetative production or growth attains its most desirable level neither will the land be fully used nor will the true (i.e. maximum possible) sustained yield be achieved.

Chapter IV

THE PRODUCTION PERIOD
OR ROTATION

The production period or rotation is the interval of time between the formation of a young crop by seeding, planting or other means and its final harvesting.

That definition is straightforward and applicable to both agriculture and forestry, to lettuces or trees, except for the difference in time involved. The great length of the production period in forestry certainly complicates the making of decisions on policy and finance and planning in general. It differentiates forestry from other uses of natural resources.

Standards of ripeness

But there are other factors peculiar to forestry, factors related both to the way trees are grown as a crop and the variable standards of ripeness applicable to trees. Trees satisfy different demands at different times and different sizes whereas an agricultural crop is grown for one purpose for which it is ripe at a definite moment of its life which can be predicted within close limits of time and decided by inspection. What standards of ripeness should be applied to trees? Size, age, vigour and rate of growth or a combination of those things?

Methods of growing trees

The way a crop of trees is grown will affect the mode of its harvesting. In broad terms trees can be grown in two different ways. In the one way the trees, whether all of one species or a mixture, are grown in uniform stands so that in any one stand the trees all have the same age. In such an even-aged forest there are, therefore, a series of separate stands each containing trees of one age but together having trees of all ages. In the other way trees of all ages are mixed together so that in each small unit, such as an acre or two, of the forest there are trees of all ages between, for example, one and a hundred years.

In the latter uneven-aged forest, trees are selected individually on their merits for felling and removal. The age of any tree is, of course, not known. So the selection of a tree for felling will depend firstly (but not necessarily primarily) on an assessment of its combined qualities of size, vigour and fitness for the market, secondly on the need to adjust the proportion of different sizes and thirdly on silvicultural principles of removing poor stems to favour better ones and to attain the desired density of canopy.

Clearly a silvicultural system of that kind is greatly flexible and enables the forester to adapt his felling from place to place to suit different rates of growth caused by variations in site or species. Size, vigour and form of individual trees are the combined criteria of felling—but not age which is unknown. Moreover the forest is perpetual and never suffers a complete clearance of trees from any piece of land but only repeated thinnings. There is no final harvesting and no production period as defined above. But one could say that its production period is equal to that of the average age of the larger trees removed—the exploitable age, i.e. the age at which a tree (or crop) attains the size required to fulfil the objects of management (Anon. B.C.F.T. 1953).

In an even-aged forest, on the other hand, whole stands of trees occupying a sizeable unit of land are felled at once (or during a comparatively short time) when they are deemed to be ready for felling. There is a clear production period or 'rotation' which may be planned in advance to give a harvest that satisfies the objects of management. The rotation has accordingly been defined as 'the planned number of years that elapses between the formation and felling of a stand' (B.C.F.T. 1953).

Superficially examined a planned rotation seems to be an admirable thing as it should facilitate planning and organization of work. But in fact there are difficulties even if size of timber alone is the object of management. Rates of growth will vary with site variations even if only one species is grown. So some stands will reach the desired size sooner than others. If several species are grown, even if each species is segregated and grown pure, the variety of times taken to reach the size wanted will be even greater. Moreover, accidents will happen—windblow, disease or fire—making it desirable to cut a stand sooner than had been planned.

If profit is the object of management the same difficulties arise. The degree of profit is affected by a combination of the factors of the rotation adopted, rates of growth, the price gradient for timber sizes and the costs of growing it. The implication of those factors on profits will be studied later. But to achieve the desired profit stands will obviously have to be felled finally at various times dependent on their rates of growth.

The removal age and its implications

Even if a definite rotation is planned, many stands will be felled at ages differing from that rotation. That felling age is termed the 'removal age'. In effect, the rotation used will equal the mean of the removal ages of all stands.

If stands are subjected to different removal ages according to their rates of growth and times taken to achieve the purpose of management it is apparent that those portions of a stand which perform better may be felled and removed before other portions which perform less well. If that is not done some parts of the stand will not fully satisfy their purpose. Accidents of wind and disease or fire will also encourage earlier felling of some parts of a stand. So there will always be a tendency to break up originally uniform stands into smaller fragments, a tendency intensified by marked variations in site (Osmaston, 1962).

An even-aged forest whose stands have been fragmented in that way until individual stands become very small will begin to approach the structure of an uneven-aged forest. Stands will be reduced to groups and be tended and harvested as groups of, perhaps, only several mature trees in contrast to the tending and harvesting of individual trees of the fully uneven-aged or true selection forest. The two contrasted methods of working forests merge into each other.

Theoretically it can be argued (Knuchel, 1953) that production efficiency increases as stands get smaller and smaller—at least for the more light tolerant species. By paying separate attention to smaller and smaller units, ultimately each tree, each unit can be more appropriately tended and finally harvested in harmony with the properties of the site and current performance of the trees on it. When the structure of the true, individual tree selection forest is attained treatment becomes almost infinitely

3

flexible and adaptable to conditions. Each tree and the ground it occupies become the unit of treatment.

But the smaller the unit of treatment the more intense is the management thereby demanding greater complications and skill in felling and extracting timber without damage to standing trees, and in managerial control. It is also essential that the species used are fully suited to the method. The comparative simplicity of even-aged forests and their aptness for a definite, planned organization in space and time persuaded the early foresters of 150 and more years ago to adopt them as the ideal type. Those advantages still obtain and are even enhanced by high labour costs and the mechanization of modern equipment which succeeds best in concentrated rather than dispersed application.

Therefore rotations of even-aged forests have been and still are a basic concept in forestry and it is necessary to examine the different kinds of rotation that may be used.

ROTATIONS

1. *Length of rotation*

The kind of rotation will depend on the objects of management, but the length of rotation, of whatever kind, will depend on the interaction of several factors, namely:

   (i) The rate of growth which varies with:
      (*a*) The species grown.
      (*b*) The soil and other site factors such as climate, topography and water supplies; in other words the site fertility.
      (*c*) The intensity of thinning.
  (ii) The inherent characteristics of the species such as its natural span of life, its age of seed production, the age when its rate of growth culminates and the age at which the quality of its timber is most desirable or begins to fail.
 (iii) The response of the soil such as deterioration or change in character after frequent exposures.
  (iv) Economics dependent on a combination of the elements of cost, prices of different sizes of wood and the times taken by the trees to attain those sizes.

Some of those factors may affect the policy of management. The demand on the soil and its consequent exhaustion caused by short coppice rotations and the removal of all the produce grown, including leafy shoots, may make sustained coppice working impossible without applying expensive fertilizers. Climate and topography may necessitate long, protective rotations or the adoption of uneven-aged forestry that uses an exploitable age and no rotation, whatever the economics may be.

It is also clear, as has been stressed before, that the inevitable combination of factors will require different rotations in different parts of a forest and thereby complicate management.

2. The various kinds of rotation

The rotations can be classified into several types to suit various purposes of management. Those usually recognized are given below. The definitions given are those of the British Commonwealth Forest Terminology (Anon. 1953) unless it is stated that they have been modified.

(i) The physical rotation. 'The rotation that coincides with the natural lease of life of a species on a given site.' This rotation has importance only in gardens, parks or protection forest and possibly amenity forest. Another interpretation of the physical rotation is the age to which trees remain sound or produce viable seed. None of them affect economic forestry.

(ii) The silvicultural rotation. 'The rotation through which a species retains satisfactory vigour of growth and reproduction on a given site.' As old trees may still retain considerable vigour and reproductive powers when the canopy as a whole opens out to allow weeds to grow and prevent regeneration the definition might be extended by adding the words 'and maintains conditions satisfactory for regenerations'.

Silvicultural rotations are not only long but also have wide limits so that they may obtain over a period of thirty, fifty or more years. They are, therefore, somewhat vague and may be used in combination with other rotations such as the technical rotation. Silvicultural rotations may be useful in forests managed primarily for aesthetic purposes of amenity where large, old trees with accompanying natural regeneration are advantages.

(iii) The technical rotation. 'The rotation under which a species yields the most material of a specified size or specification for a special use' (B.C.F.T. modified). This rotation is often used and examples are many, such as hop poles in sweet chestnut coppice, telegraph poles in conifers or large, slow grown oak for conversion to veneers.

The technical rotation may therefore be long or short and may be fairly precise—e.g. within a year or so of twelve years for sweet chestnut coppice for hop poles. One can note that by use of a long technical rotation such as that intended to produce large timber for a particular market, much other smaller timber will also be provided from thinnings during the rotation. One may also note that an even-aged stand always contains a large range of tree sizes about its mean size. If a stand has a mean breast height diameter (DBH) of 25 inches, the range of DBH may easily be from 18 to 32 inches; half the number of trees being less than 25 inches. So if timber 25 inches DBH and over is wanted, the technical rotation should probably be longer than that giving a mean 25 inch DBH in the stand.

The technical rotation is used particularly by industrial companies which own forest for the purpose of supplying timber for their plant, such as paper pulp companies. But for them and others, if size of material is not paramount, the rotation of highest volume production may have great attractions.

(iv) *The rotation of the greatest volume production*. 'The rotation that yields the greatest annual quantity of material.' That quantity (which is usually assessed as the volume of wood above a minimum thickness) naturally includes material from all thinnings as well as the final volume felled at the end of the rotation. It is a very important rotation and is frequently used.

The length of the rotation will obviously coincide with the year when the average rate of growth or volume increment per acre reaches a maximum. It is therefore necessary to consider the volume increments of stands.

Volume increment

The current annual increment (CAI) of a stand is the increase in its volume during a year. The periodic mean annual increment (PMAI) is the increase in volume during a period of years. As it is not easy to measure precisely the CAI it is usual to measure the PMAI over a short period of five or possibly ten years and assume that it is the same as the CAI.

The mean annual increment (MAI) is the mean of all CAIs to date, and can be expressed by the formula $MAI = (Y_r + \Sigma T_r)/R$ where Y_r is the volume of the final felling, ΣT_r is the sum of the volume of all thinnings and R is the rotation.

As shown in Fig. 1 the CAI of a stand increases slowly in extreme youth, then accelerates and then slackens until the

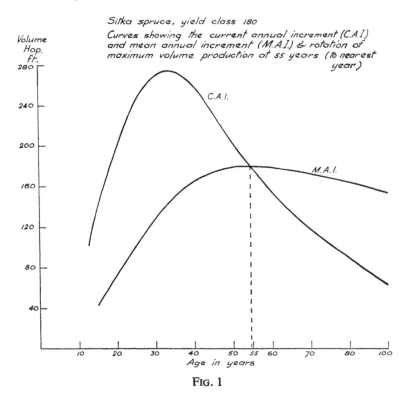

Sitka spruce, yield class 180

Curves showing the current annual increment (C.A.I.) and mean annual increment (M.A.I.) & rotation of maximum volume production at 55 years (to nearest year)

FIG. 1

CAI culminates after which it progressively falls. With even greater age, as growth slows and is accompanied by the death of trees in the stand, the CAI will become negative.

The MAI, being the mean of all CAIs to date, has a more regular and restrained progress. At first, owing to the acceleration of the CAI from nil, the MAI is always less than the CAI but continues to rise for a longer time until the two become equal. Then the MAI will also fall but, being still influenced by the earlier high values of the CAI, will always remain the greater and will never reach zero. So the moment when the CAI and

MAI are equal is the moment when the MAI achieves its maximum, and that is the rotation of the greatest volume production.

Table 5

MEAN ANNUAL INCREMENTS OF SOME CONIFEROUS SPECIES IN BRITAIN

(Taken from the British Yield Tables, 1966)

Species and (Yield class)		MAI in hop. ft./acre at stated ages in years										
		30	35	40	45	50	55	60	65	70	75	80
Scots pine	(80)	34	43	52	59	65	70	74	77	78	79	80
	(120)	72	84	95	103	110	114	117	119	120	120	119
	(160)	114	128	139	147	153	157	159	160	160	159	157
Jap. larch	(80)	70	76	79	80	80	79	78	77	75	73	71
	(120)	112	117	120	120	119	117	117	115	111	109	106
	(160)	154	159	160	159	158	156	153	150	147	144	141
Douglas fir	(140)	103	117	127	134	138	140	140	139	137	135	132
	(200)	164	181	190	196	199	200	199	197	194	190	186
	(260)	223	242	253	258	260	259	257	253	249	244	238
Sitka spruce	(140)	91	108	121	131	137	140	140	139	137	134	131
	(200)	153	174	188	196	200	200	198	194	190	185	180
	(280)	247	266	277	280	279	276	270	263	256	248	240
Norway spruce	(140)	77	95	109	119	127	133	136	138	140	140	139
	(200)	131	154	171	182	190	196	199	200	200	199	197
	(240)	167	193	212	224	233	237	239	240	239	236	233

An important point to notice is that for any one species change in quality (i.e. yield class) does not materially affect the rotation of highest volume production. Thus for Douglas fir a rotation of fifty-five years provides the maximum volume for both yield classes 140 and 200 and only 1 hoppus foot less than the maximum for yield class 260. Nevertheless the rotations for maximum volume are slightly longer for poorer yield classes than better ones for all species, and occur earlier for pronounced light demanders such as Japanese larch.

(v) The rotation of highest income. 'The rotation that yields the highest average net income' (BCFT modified). This rotation is also known as the rotation of the highest forest rental. It should be noted that no regard is paid to the amount of capital used in earning the income.

The average net annual income or rental (F_r) obtained from a stand of trees is expressed by the formula:

$$F_r = \frac{Y_r + \Sigma T_r - C - \Sigma_e}{R}$$

where Y_r is the net money value of the final felling, T_r is the net value of all thinnings during the rotation, C is the cost of formation of the stand, e is the annual cost of administration during the rotation and R is the rotation.

For calculating its length this rotation is similar to that of the highest volume production. Thus the net volume production (i.e. the MAI) also $= Y_r + \Sigma T_r / R$ where Y_r and ΣT_r are the *volumes* of the final and thinning yields instead of their *values* as used in the net income formula. To use the net income formula, therefore, it is necessary to multiply the volume yields expected by the net prices expected for the timber; in other words to use a money yield table instead of a volume yield table, and, in addition, subtract costs of formation and maintenance.

It is apparent, therefore, that the two rotations will be much the same length unless there is an appreciable increase of price for larger sized timber which is, in fact, common. If that price gradient is marked then the rotation of highest net income will be comparatively long. Also, if there is a special size of timber that fetches a particularly high price the rotation that provides that size may be the rotation of highest income and possibly coincide with the technical rotation.

(vi) The financial rotation. The BCFT defines this rotation as that which is 'determined by financial considerations, e.g. that yielding the highest interest'. A simpler but nevertheless wide definition is that given by Hiley (1956) 'the rotation which is most profitable'. In the current context the words 'most profitable' must imply the greatest financial profit or advantage which requires an assessment of the monetary gain derived from the investment made in the activity. But even that concept can be combined in more than one way in forestry. Two approaches may be considered:

(a) The expectation value of the land, i.e. the value based on the net income which it is expected will be obtained from it and calculated at a selected rate of interest.

(b) The financial yield—i.e. the rate of interest, or mean annual forest per cent, which the forest enterprise yields on the money invested in it.

The land expectation value

In forestry this value is often known as the soil expectation value S_e, a term used by Schlich (1925 and earlier). But as pointed out by Hiley (1956) the soil is obviously not the only factor that influences the growth, and income or costs of a tree crop. Many other site factors affect the matter such as topo-graphy and extraction costs, nearness to markets, local climate and so on. A better term is land expectation value.

If a piece of land is expected to provide a continual net income of £x yearly then that land can be valued at a sum which at an acceptable rate of interest, such as that given by a long dated government stock or other investment of similar security, gives the same yearly income of £x. That value is known as its expectation value, S_e. Expressed by formula when

$$p = \text{the rate of interest, } S_e = \frac{x}{0 \cdot 0_p}.$$

But if the piece of land produces income periodically instead of yearly, such as coppice forest cut every tenth year, the present discounted value of that return $= Y_r/1 \cdot 0p^r - 1$ where Y_r is the net periodic income produced every rth year for ever.

Consequently a formula can be derived to calculate the expectation value of land by discounting to the present all forecasted future net income whether collected yearly or at regular intervals and subtracting from this sum the discounted forecasted future expenses calculated in the same way. Such a formula, usually known as the Faustmann formula, is shown below:

$$S_e = \frac{Y_r + T_a 1 \cdot 0p^{r-a} + \ldots T_q 1 \cdot 0p^{r-q} - C.1.0p^r}{1 \cdot 0p^r - 1} - E$$

where Y^r is the net value of the final felling made in the year r at the end of the rotation, $T_a \ldots T_q$ are the net values of the several thinnings made in the years r- a ... r—q, C is the cost of planting the forest crop at the beginning of the rotation, p is the selected rate of interest and $E = e/0 \cdot 0p$ where e = the sum of all annual expenses.

The formula depends on the assumption that each item of income or cost recurs at definite and constant intervals of time for ever and is the same constant figure at each recurrence. Each item of the formula thus becomes the sum of an infinite series of discounted costs. Thus the same final net yield, Y_r, is received at the end of r, 2r, 3r years, etc., for ever and the sum of the infinite series of discounted values of $Y_r/1 \cdot 0p^r$ equals $Y_r/1 \cdot 0p^r - 1$.

In effect the calculation is one of computing the net discounted revenue (NDR) and the true net discounted sequence is often used instead of land expectation value.

The value of S_e (or NDR) will clearly vary with the rate of interest, p, and the length of rotation, r. The equation cannot be solved for S_e if both p and r are unknown. But if a rate of interest is chosen and accepted, the most profitable rotation will be that which gives the maximum value for S_e (or NDR) and the crop of trees on the land will then yield the selected rate of interest on the money invested on the costs of formation of the plantation, its maintenance and harvesting.

Thus if the money for all expenses, purchase of the land cost, of planting and annual costs of administration and main-tenance have to be borrowed from a bank at 5 per cent interest, it may be found that the most profitable rotation is fifty years which gives a value of £45 per acre for the expectation value. If the land is then sold at the same figure as it was bought for and the money repaid to the bank and if all the income received is also paid to the bank, the debt will then be exactly discharged if the price paid for the land was £45 per acre. The bank will have profited by 5 per cent on its loan, i.e. its investment.

But if the cost of the land, S_c, was only £20 the entrepreneur will make a profit of £25, that is to say, that the profit (P) can be defined to equal $S_e - S_c$, namely the difference between expec-tation and cost values. Thus, using the same symbols as before:

The present value of all net returns

$$= \frac{Y_r + T_a \times 1 \cdot 0p^{r-a} + \ldots Tq \times 1 \cdot 0p^{r-q}}{1 \cdot 0p^r - 1}$$

The present value of all expenses $= S_c + C + E$

But C as the discounted value of the cost of formation re-peated every r years $= \dfrac{C \times 1 \cdot 0p^r}{1 \cdot 0p^r - 1}$

Hence the profit P, which is the difference between the present values of net returns and expenses

$$= \frac{Y_r + T_a \times 1 \cdot 0p^{r-a} + \ldots T_q \times 1 \cdot 0p^{r-q}}{1 \cdot 0p^r - 1} - S_c - C - E$$

which can be rewritten as:

$$P = \left(\frac{Y_r + T_a \times 1 \cdot 0p^{r-a} + \ldots Tq \times 1 \cdot 0p^{r-q} - C \times 1 \cdot 0p^r}{1 \cdot 0p^r - 1} - E \right) - S_c$$

$$= S_e - S_c$$

3*

The financial yield

As pointed out above the value of S_e depends on both p and r. As the formula is not soluble for both unknowns the values for S_e have to be calculated for different rotations for a range of chosen values of p. A series of curves, each for a different value of p can then be drawn to show the values of S_e for different rotations as shown in Fig. 2 (after Hiley (1956)).

An indicator line, CC, can then be drawn through the peaks or culminating points of each curve to show the year when the S_e culminates for a selected rate of interest—i.e. the rotation of maximum NDR.

But the rotation at which the highest rate of income is obtained is the point where a 'p' curve culminates at the point where $S_e - O$, i.e. where the indicator line cuts the x axis.[1] That will be the financial rotation, namely that at which the plantation earns the highest rate of interest.

It can be seen from the figure that the financial rotation, i.e. that which gives the highest financial yield, is shown by the point where the indicator line CC cuts the x axis where $S_e = O$, a rotation of sixty years. The dotted curve shows the approximate curve for a rate of interest of about 3·9 per cent. But the figure also shows that the rotation which gives the highest expectation value (or NDR) differs according to the rate of interest used, namely about sixty-eight years for 3 per cent, sixty-three years for $3\frac{1}{2}$ per cent and fifty-six years for 5 per cent. The need to distinguish between the rotation of the highest financial yield and the rotation at which the NDR is a maximum is clear.

Weaknesses of the Faustmann formula

For estimating financial rotations the Faustmann formula has several weaknesses such as those given below:

(i) The formula is only applicable to even-aged forestry

---

[1] The actual cost or market value of the land is excluded from the calculation. It must be included, e.g. as part of the cost of formation (C), when deciding whether the land is to be used for forestry or some other purpose; the financial yields expected from forestry and other uses of the land can then be correctly compared. But having decided to use the land for forestry the value of the land does not affect choice of the financial rotation which gives the highest financial yield.

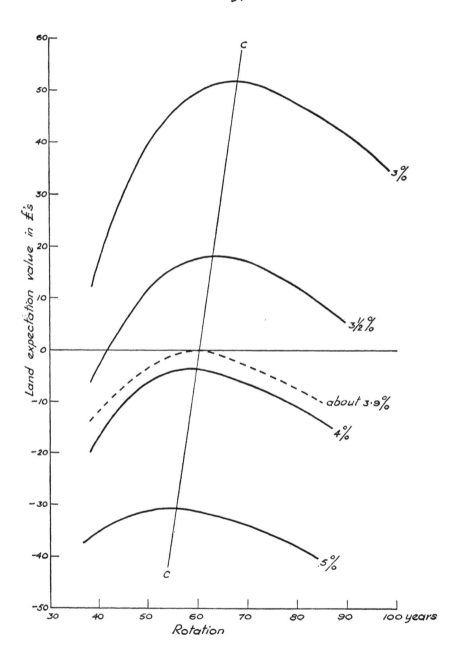

Indicator graph Norway spruce Q.Cl.II,
moderate Thinning (British yield tables 1956)
(after Hiley)

FIG. 2

where there are definite times for formation and regeneration, for periodic and final harvests from particular pieces of land. It cannot be used for uneven-aged forests worked by selection fellings with no rotation. For them a full financial assessment of the profit over and above a forecast of net income is difficult. But that net income, expressed without consideration of accumulated compound interest, can be related to the value of the standing growing stock and shown as a percentage of it. The rise and fall of the value of the growing stock and the current income as a percent of the capital are easily assessed as a simple interest problem.

(ii) The formula depends on the recurrence of the same money values at regular intervals whereas they may change in value and not recur at exactly the same intervals.

(iii) No allowance is made for irregularly occurring expenses (which must occasionally occur) except as an average figure debited to annual expenses but forecasted regular expenses (e.g. pruning) can easily be included at fixed rates and times.

(iv) To keep the formula simple it is assumed that the whole cost of formation occurs at the beginning of the first year whereas in fact there will usually be some delay and some costs of formation (e.g. weeding and beating up failures) will occur two or more years later. Also the annual expenses, e, are assumed to be paid at the end of each year but, in fact, will be made during the year. However, these errors do not greatly affect the result.

(v) For estimating the expectation value of the land or NDR a definite rate of interest which has a decisive effect on the result has to be chosen. But it is not easy to choose the correct rate of interest. If the money has to be borrowed, e.g. from a bank, then the rate of interest charged on the loan can be fairly used. If the owner himself provides the money used he can charge the rate of interest that he would have got by investing the money in an alternative project of similar promise and security. But it is not easy to judge security and comparability of alternative projects so that settlement of the rate of interest in that way is not easy.

(vi) Intangible benefits of forestry are not included in the formula but could be incorporated (e.g. as a deduction from the annual expenses or by acceptance of a lower rate of interest) if their annual values in terms of money are known. But they are seldom known.

Consequently, financial rotations, either those based on the maximum land expectation value (net discounted revenue) or those that give the highest financial yield (internal rate of return), can be approximate only and liable to change as conditions change. Nevertheless calculation of the financial rotation is a valuable, even an essential guide for planning the organization of a forest growing stock, namely in:

(i) Provision of the approximate rotation needed for economic management so that yields and financial prospects and programme of work can be forecast and arranged.

(ii) Comparison of the economic prospects to be expected from growing different species so that decisions of choice of species can be made. The rate of interest used does not affect the comparison, only the amounts of difference.

(iii) Estimation of the economic disadvantage or loss caused by adopting some rotation other than the financial one.

Having decided on an approximate financial rotation for the purpose of general planning to obtain the best financial results the actual removal age for finally felling a stand is then calculated and decided quite separately and individually for each stand according to its condition in relation to the prospects of its successor and current market and other factors (subject to any other constraints that may apply). That removal age can be decided by using the Faustmann formula, with current prices and costs, to calculate the year when the combined NDR of the stand and its successor culminates. (See Appendix III(c) for an example.)

For a full exposition of the Faustmann formula and financial rotation works on forest valuation and economics should be consulted (e.g. Hiley, 1956). It is, however, worth noting here that financial rotations are comparatively short (the higher the rate of interest the shorter they are) and that income received early in a rotation, e.g. from thinnings, has a substantial effect on the length of rotation especially at a high rate of interest. Heavy thinnings lengthen the rotation. A longer rotation combined with the faster individual growth of the fewer stems left after thinning, enable larger timber to be grown to financial advantage than would otherwise be possible.

Discussion on rotations

There are two aspects of rotations to consider—which kind of rotation should be chosen and how that rotation should be applied in practice.

1. *Choice of the kind of rotation*

For considering the choice of the most suitable types of rotation we may divide rotations into three main groups which satisfy three different purposes, namely:

(*a*) Rotations to control the supply of certain services, i.e. the silvicultural and physical rotations.

(*b*) Rotations to control the output of material forest goods in form or quantity—i.e. the technical and maximum volume rotations.

(*c*) Rotations to control the monetary returns, i.e. the rotations of maximum gross or net income and the financial rotation.

They will be discussed separately.

(*a*) *Service rotations.* As explained earlier in this chapter the physical and silvicultural rotations are both very long but may be chosen in parks and gardens, in forests used for protecting soil stability and in forests managed for beauty and recreation. But for such purposes it may be better to have irregular forest worked by selective fellings. There is then no rotation but the exploitable age may approximate the silvicultural or even the physical rotation.

If the objects of management require a very long technical rotation which is longer than the silvicultural rotation (very unlikely) the latter might well have to be chosen to maintain productivity.

(*b*) *Rotations controlling the material output.* In certain conditions the purpose of the owner or owning body may require from the forest the production of a particular product or the maximum quantity of woody produce rather than more favourable monetary returns. The technical or maximum volume rotation may then be attractive. A few examples may be quoted.

In an agricultural community in an undeveloped region the very existence of the people may depend on adequate supplies of firewood, fencing material, poles and small timber for house-building, all obtainable from one nearby forest. If the owning body of the forest concerned is the government or local community it will then probably choose the rotation of maximum volume production provided that it is long enough to produce the sizes of material wanted. If the rotation is too short for those sizes the technical rotation becomes attractive. In any event the owning body is concerned with the welfare of the people and not monetary gain from the forest.

In some circumstances a community (or forest owner) may depend for their livelihood on the supply of a forest product to a close specification, for example small posts for vineyards. The forest needed to satisfy the demand would then be worked in the (very short) technical rotation unless there were to be enough forest to supply the essential posts from thinnings and the tops of larger trees grown primarily for another purpose.

An industrial firm owning both forest and an industrial plant dependent on forest produce will be interested in magnifying the combined profits of the plant and forest. If the specification of the essential forest produce is narrow the technical rotation is likely to be chosen particularly if that technical rotation has to be long. If, however, the specification is for small produce which can be cut from thinnings of larger trees then a rotation longer than the technical rotation might be chosen if it were more profitable. If the specification is a wide one then the rotation of maximum volume production might be chosen since it might enable the industrial plant and its profits to be expanded.

The need to improve its international balance of payments might influence a government in choice of rotation. Thus, if a country enjoys an export trade in large timber of veneer quality it may wish to allot certain forests to the production of that timber and use the technical rotation thereby needed even if it was less profitable for the individual forests. The nation's general advantage would be gained at the expense of localized loss of forest profits.

Those examples of choice of rotation to control the output of forest produce emphasizes the need to examine their advantages relative to the advantages given by rotations chosen to control monetary income. The monetary or financial advantage is usually the best means for distinguishing the best mode of managing a concern unless that concern has to supply material needed for the common welfare or for the consumption of the firm itself. The goods so supplied then, in effect, become a service, which nevertheless has to be paid for probably by a wider community, e.g. by the general taxpayer or by the industrial plant to obtain its raw material.

(c) *Rotations to control monetary returns.* Controversy becomes marked when considering the advantages of financial rotations and those providing the highest net income and highest gross income.

When considering those three types of rotation one must bear in mind that usually the growing stock and capital that has been or will be invested are least for the financial rotation, greater for the highest net income and greatest for the highest gross income; the rotation increase in the same order, the financial rotation being quite low perhaps between seventy-five and thirty or fewer years.

The status of the owning body will certainly affect their outlooks on these three rotations. For State ownership it can be maintained (for example by Knuchel, 1953) that to provide the greatest good for the nation as a whole, the paramount object should be to produce the greatest value of raw material from each unit of land. By so doing the maximum benefit will be obtained by all wood-using industries, by other consumers and by all employed in the forest and those industries; the utmost wealth, in other words, flows into the community from the source of a constantly renewable and widely used raw material. Gross income measures the value output, but ignores the costs of production. So by providing the utmost output in value the forest itself may work at a low profit or even at a financial loss; but the nation as a whole obtains the greatest possible value of raw material. Particular beneficiaries are consumers of wood near the forest, wood-using industries and all employed in them and the forest.

Use of the rotation of greatest gross income is somewhat similar to a decision to extract the last ton of coal from a mine irrespective of the cost. The gross income rotation seems justifiable only if wood is in very short supply, cannot be replaced by substitutes, cannot be imported without undesirable effects on the balance of foreign payments or if limited amounts of forest land are available. To those provisos might be added the possession of ample capital without which it might be preferable to reduce the forest capital and spend less on it in order to provide money for investment in other more profitable enterprises. Another consequence of adopting this rotation is that it would in effect subsidize the wood-using industries and the community living near and on the forests at the expense of the general taxpayer. The advantages of that are debatable.

It seems very doubtful whether any State would find the rotation of gross income desirable, involving as it does some extra wealth of raw material produced at high cost. No private owner could contemplate its use.

The rotation of highest net income or forest rental has undoubted attractions. It is widely used in long established and ably managed forests on the continent of Europe as in Switzerland and Germany. It expresses the profit and loss account of a forest. As this rotation provides the greatest net income attainable from forest land (with the species used) it must interest any owner, whether a private individual or the State. By introducing costs, although excluding capital invested, the greatest amount of raw material obtainable without unreasonable current expenditure, is achieved—the land is being fully used as a producer of income.

But the costs of capital are ignored. If the capital in the shape of growing stock has already been accumulated, especially if that has been done at no cost to the owner (by inheritance or initial use of virgin forest) the rotation is attractive. But nevertheless, by reducing the growing stock capital to that which provides the best financial yield, the surplus capital so realized could be invested in other useful enterprises. So use of this rotation implies that either the owner has ample capital or there is a lack of alternative investment. If on the other hand the comparatively large growing stock needed for this rotation does not yet exist and has to be accumulated, such as by afforestation of bare land, the cost of capital becomes very real; the growing stock has to be produced and paid for. In effect the rotation then entails current sacrifice, by waiting for the capital to pay results, for the benefit of posterity; a capital larger than necessary is accumulated by current savings for future good. Again the possession of ample capital is needed.

If the forest is already there, as a self-supporting enterprise, income is not interrupted; thinnings and final fellings occur regularly and annually. So costs which could or should be classed as capital expenditure, such as improvement of the road system, can be subtracted from income so that no interest is payable in cash or accumulates. There is only a reduction of current net income and no debts to pay. It can then also be maintained that the longer rotation and greater volume and value of the growing stock than is financially necessary or desirable provides an insurance for future emergencies or change in the demand pattern. It is easier and less costly to reduce rotations than to increase them. Insurance may be wise but, again, is current sacrifice by savings for future security desirable? Is it or is it not worth while? Who gains?

Financial Rotations, both for the best financial yield and for the highest expectation value for the soil (or land), are comparatively short and the capital invested, exemplified by the volume of the growing stock, comparatively small. That causes perhaps, a psychological reaction from many foresters who instinctively prefer stands of larger trees to smaller ones. It is difficult to deny that there is a greater pride of possession and management of a forest containing large trees than of a forest that contains only comparatively small ones. If to that fondness of having large trees is added a large net income the satisfaction is so much the greater. There is, therefore, a tendency for more attempts to be made to discredit the benefits of best financial yields than to approve them.

Few, however, try to refute the value of the financial rotation either when afforesting bare soil, for which purpose in particular the Faustmann formula was designed, for comparing the performance of different species or different uses of the land. But approval of those uses of Faustmann's formula is qualified by the difficulty always experienced in selecting the rate of interest. Thus H. Arthur Meyer *et alii* (1961) state with reference to a forest managed as a going forest enterprise 'interest rates compounded over periods of more than twenty to thirty years soon leads to unreasonable results, at least when applying interest rates larger than 3 or 4 per cent'. Yet if those are the facts of life why are they unreasonable? How can they be ignored? Or why? The only acceptable circumstances for ignoring them, as stated earlier, is when the forest capital has cost the owner nothing (e.g. by inheritance or ownership of virgin State forest). Even then realization of part of the capital by reducing the growing stock to that financially justified enables other investment to be made—in afforesting more land or in some other more lucrative enterprise. Is it not exactly that which in fact has happened when virgin forests have been exploited, except that so often has no part of the capital realized been reinvested in the forest land?

It is of course true that there are considerations that influence modifications of the financial rotation. If its use entails so short a rotation that the fertility of the soil is affected either the rotation must be lengthened to that which ensures sustained fertility or a combination of soil maintenance and improvement by draining, fertilizing and other cultivation must be incorporated at an additional cost in the management, as in agriculture. Also the rate of interest used may be lowered to give longer rotations in order to provide insurance against emergency or to better amenity or other demand considerations. But to do so does not invalidate the use of financial rotations or the rate of interest itself but rather indicates the cost (and value) of providing the considerations.

There is always the argument that land should not be wasted particularly when the present explosion of population must eventually make all land scarce. That argument would say that it is better to get only 1 per cent on capital if that land is thereby made or kept productive of raw material. The argument is valid if the prospects are so black or time is so short that the capital involved cannot be better invested elsewhere. But, if that critical shortage of land or raw material does threaten, prices of timber will then rise and make investment in forests more attractive; previous marginal land will become profitable.

Foresters may fear that financial rotations in managed forests will steadily eliminate large trees which will then be non-existent for demands which may continue or re-appear for them. There is truth in that contention. The virgin forests too will disappear and it is unlikely that the gigantic trees in them will ever be replaced at least in the civilization which is now developing. Very large trees, so expensive and slow to grow, will be found only in ecological preserves. But it must not be forgotten that financial rotations vary not only with the rate of interest but with the price gradient for timber. Two factors are important there. High prices for small timber (e.g. pulp wood) provide good profits from earlier thinnings; they tend to lengthen the financial rotation materially, as shown in Britain today where financial rotations (based on maximum NDR) for selected conifers are some fifty to sixty years when using 5 per cent interest. Then, when a shortage of big timber occurs the prices for it will rise, perhaps markedly, an event which will also lengthen the rotation.

Financial rotations, therefore, change with demand. But, owing to the slow growth of trees to harvestable sizes, adjustment of supply to demand is slow especially to larger sizes for which waiting is inevitable. There is therefore a strong reason for adopting rotations that are rather longer than the financial rotation calculated at current prices and costs. That adjustment can be made either by subjective, arbitrary decision or more logically as is done by the Forestry Commission in Britain, by analysing trends of future demands and prices. If the price of timber is expected to rise faster than that of other commodities then income can be discounted at a lower rate than costs. If, in addition, the value of money is expected to fall then the rate of fall can also be subtracted from the rate of interest.

In conclusion of this discussion on choice of rotation it is worth referring to and emphasizing the advice given long ago by Schlich (1925), namely, that the financial rotation should always be determined, even though it may necessarily be approximate, so that the financial loss caused by the adoption of another rotation, considered to be better for satisfying the objects of management, is determined and made clear. As pointed out by Hiley (1956) the financial yield is the rate at which forestry *creates* wealth. As there is always competition for capital, capital should always be applied to those uses which create wealth most rapidly. In forestry the rotation of maximum financial yield creates wealth more rapidly than any other rotation and should usually be used. It compares the results of forestry with the results of other undertakings and, in forestry itself, shows the cost of adopting other rotations. As those trends apply to both social and private wealth the best financial yield should be the aim of the State as much as the private owner although the cost of intangible benefits may be a more acceptable charge on the State than on private ownership.

2. *Application of rotations*

The adoption of a rotation in even-aged forestry enables the forester to organize in space and time a regular sequence of maturing timber and consequent operations. That organization may be rigid so that stands are felled when they reach the age of the predetermined rotation and tended at regular intervals in their lives. Simplicity of management is the great advantage and outweighs, it is hoped, the sacrifices inevitably experienced in attaining the initial design which is sure to require some stands to be felled either before or after they reach rotation age. But, even when the designed structure and simplicity are achieved, the rigidity allows no easy adjustment for accidents, market and price changes or variety in growth of individual stands.

Greater flexibility is usually desirable and leads to separate treatment for individuals stands. Each stand, being treated on its own merits, will then be finally felled, removed and regenerated when its removal best suits the objects of management. The best possible designed use of the land is then achieved. The removal age may often differ from the planned rotation but the mean of all removal ages should equal that rotation which, however, may have to be changed as the result of experience if the mean removal age is persistently different.

We see, therefore, that the rotation guides the main organization and broad planning of fellings within which removal ages are decided to suit the current state of each stand, accidents and, to some extent, markets.

Decision on removal ages, when stands should be felled on their merits, are not always easy to make. For stands damaged by wind or fire it may be fairly easy to judge whether the volume left can provide an acceptable increment per acre. If the object of management is the provision of a particular type or size of tree it may also be easy to assess the time of felling. If the object is maximum volume per acre per annum the removal age can be decided fairly precisely by the point when the CAI of a stand falls below a predetermined figure for the quality or yield class of the stand.

If the best financial results are wanted decisions on the removal age are not so straightforward, largely because the calculations are liable to be tedious and involve forecasts not only of increments but also of costs and prices and the times when they occur. It is not the financial rotation itself that matters but the present performance of each stand, in income at once available or expected later, in relation to the expected performance of a successor crop.

Partly because the method simplifies comparison of several, different courses of action, the British Forestry Commission uses the device of calculating the net discounted revenue (NDR) for each £100 invested as their standard of profitability (Working Plan Code, 1965). The income and expenditure are calculated and discounted separately and the total discounted expenditure is subtracted from the total discounted income to give the NDR which is expressed so that NDR obtainable from each £100 invested (i.e. of the discounted expenditure) and not as the NDR per acre or unit of land. By giving standard weighting factors to certain variables such as accessibility to markets, species and road intensities, and by the use of yield class and thinning production tables, a standard procedure for calculating and using NDR has been evolved. It is, therefore, comparatively simple to calculate the NDR for any stand and also for a variety of successive crops, whether the stand is felled and regenerated at once or later. The course of action that gives the highest NDR will be the most profitable.

For example, consider a stand whose final felling now would produce a net income of £200. Alternatively the same stand might be thinned and finally felled ten, twenty or thirty years hence to give greater sums. But those sums should be discounted to give the present NDR; suppose these to be respectively £235, £240 and £230. Those figures suggest that it would be best to fell twenty years hence. But that is not the end of the study. What are the prospects of the succeeding stand, formed at the time of final felling? The NDR of that stand must also be calculated, discounted to the present and not only to the year of formation. Suppose the present values of the NDR's of successor stands formed now, ten, twenty or thirty years hence are £90, £60, £40 and £25. Then to fell and regenerate now would give a NDR of £200+£90 = £290; to fell and regenerate ten, twenty or thirty years hence would give NDR's of £295, £280 and £255 respectively. It is better to fell ten years hence.

We may conclude that the rotation, whether it is the financial, technical or any other rotation, is used mainly for broad silvi-cultural planning and organization of the growing stock rather than for deciding the actual time for final felling and removal of a stand. As we shall see later (chapter VI) the rotation is an important element in organizing forests worked under a periodic block method and may be an essential factor in formulae used for calculating the permissible yield (chapter VIII). Moreover even when no formula is used for calculating the yield, the rotation must still influence decisions on the amount of final fellings whenever sustained yields are wanted since regularity of the yield depends on the proper proportion of all age-classes which in turn depends on the length of the rotation. The rotation then becomes a constraint on choice of removal age since some stands may have to be kept standing longer or removed sooner than their actual state would otherwise require. If, however, sustained yields are not included in the objects of management the rotation becomes only a guide to treatment and an aid to long-term forecasts of future yields; decisions on the time when finally to fell and remove a stand will then depend only on its state and performance and current market and other circumstances.

As pointed out in the last chapter the need for regularity and sustention of yields from individual forests has faded with the improvement of transport, the changing pattern of demand and increase in the size and concentration of wood-processing industry. Wherever those changes apply (and they do not yet apply in many underdeveloped tropical countries) rotations, together with sustained yields, lose much of their hold on forestry thinking and make possible a more flexible and prac-tical approach to decisions or the removal of individual stands.

Bibliography

Anon (1953) 'British Commonwealth Forest Terminology.' Empire Forestry Association, London

Anon (1953) 'Working Plans Code of the British Forestry Com-mission'

Hiley, W. E. (1956) *Economics of Plantations*. Faber and Faber, London

Knuchel, H. (1953) *Planning and Control of the Managed Forest*, translated by M. L. Anderson. Oliver and Boyd, Edinburgh

Meyer, H. A., Recknagel, A. B., Stevenson, D. D., Bartoo, R. A. (1961) *Forest Management*, 2nd edition. The Ronald Press, New York

Osmaston, F. C. (1962) 'The Management of Coniferous Forests in Britain', supplement to *Forestry*, 1962

Schlich, W. (1925) *Schlich's Manual of Forestry*, Vol. III. 5th edition. Bradbury, Agnew and Co. Ltd, London

Chapter V

THE NORMAL FOREST

In forestry, as in any other undertaking, there should be an ideal state of perfection which satisfies the purpose of management to the full. Without such an ideal design, the organization of ways and means becomes confused and there is no standard by which to judge the efficiency of management. The ideal and the consequent organization of methods of action are as essential for efficiency as are clear objects of management. In forestry this ideal state of perfection is called the 'normal forest'.

Owing to the organic constitution of forests, the normal forest is perhaps more nebulous and liable to change than the ideal design of most other enterprises. But no undertaking can have an absolute normality; changes in standards of living and markets, progress in invention and techniques may change even the purpose of management and are bound to change methods and therefore the state of normality. Forestry is similarly affected, and, in addition, is exposed to fluctuations of climate and climatic or other disasters from wind, drought, fire, insect or fungal diseases, all of which can upset the degree of normality that has been achieved. There will, therefore, be not only a changing state of normality with which slow growing trees have to be adapted, but also accidental interruptions and interference. So in practice the normal forest is never achieved or, if achieved over part of the forest, is never maintained for long.

Nevertheless to have an ideal towards which to strive remains essential, on which to base action and avoid confusion and inefficient performance. But, it may be asked, if the 'normal forest' is liable to be a somewhat nebulous and unattainable concept, can it be defined sufficiently clearly and precisely for use and yet retain enough plasticity in execution?

BASIC FACTORS OF NORMALITY

From what has been said it is clear that a normal forest must be the one that best satisfies the chosen objects of management. To do that the normal forest must possess certain general attributes, namely:

(i) The specific composition and the structure or form of the forest must harmonize with the environment or factors of the locality. In other words the species grown and the methods of silviculture adopted must fully suit all peculiarities of site; only then can full growth be secured.

(ii) The growing stock of trees must be so constituted that it provides regularly the greatest possible quantity of the desired products (which naturally include intangible benefits).

This second attribute, therefore, includes a sustained yield which is usually an essential element in the objects of management, at least for a group of forests. If a sustained yield is needed only for a group of forests, the concept of normality, in this sense, may be extended to the group rather than to an individual forest or part of a forest.

Moreover, this attribute, in combination with attribute (i), implies that the chosen product must come from a species capable of happy growth in the environment.

(iii) The general organization of the forest must be appropriate for its purpose—for example it must have an adequate road system and extraction methods co-ordinated with the sales organization.

(iv) The organization of the forest into working units and the general administration must be the best possible.

Those four attributes, all of which in combination are needed for full efficiency in working a forest, are not only silvicultural but managerial and administrative. But the term 'normal forest' is usually understood to include only the silvicultural organization (attributes (i) and (ii) above) and the consequent managerial organization of silvicultural working units (from attribute (iv)). Nevertheless, it is useless to have a forest perfect in all its parts except for adequate administration and communications. A forest enterprise must be considered as a whole.

Let us now consider the narrower and classic concept of the normal forest, that is to say, its silvicultural organization, and see how it can be defined and then expressed in concrete or numerical terms.

Definition of the normal forest

Even the narrower concept of silvicultural normality must be a wide one as it should apply to all types of forest and all purposes of working. One might define the normal forest as 'that forest which has reached and maintains a practically attainable degree of perfection in all its parts for the full satisfaction of the purpose of management'. (Compare Knuchel's (1953) similar definition.)

In that definition the words 'practically attainable' have been introduced to exclude a theoretical perfection impossible ever to attain. An example of unattainable perfection is complete, ideal stocking of all stands such as the ultimate stocking given in yield tables. Those yield table figures of volume, true for selected and small units of area such as one acre, are impossible to get in practice over many acres; experience might show that in practice ideal, normal volumes should be 5 or 10 per cent, even 15 per cent less than those of the yield tables.

It is interesting to compare the definition proposed above with that given in the British Commonwealth Forest Terminology (Anon. 1953) which reads: 'A standard with which to compare an actual forest so as to bring out its deficiencies for sustained yield management; a forest which, for a given site and given objects of management, is ideally constituted as regards growing stock, age class distribution and increment, and from which annual or periodic removal of produce equal to the increment can be continued indefinitely without endangering future yields'.

It is noteworthy that the definition refers only to the growing stock and that a sustained yield is an essential part of it.

Historical background of the normal forest

The idea of normal forests began to take shape among foresters of the late eighteenth and early nineteenth centuries (e.g. G. L. Hartig in his book *Anweisung zur Holzzucht* in 1791, and later by Cotta, Hundeshagen, and others), when the principle of sustained yield took root. Nor must one forget the famous but anonymous Austrian tax collector of 1788 who, for the assessment of the value of forests, introduced the principle that forests should be capable of continuous, regular yields. Acceptance of that principle demands that the forest should be so constituted that it has trees or stands of differing ages, from youth to maturity, so that there is a steady, preferably annual sequence of maturing timber. Such a forest will have a standing volume of timber just right for that constitution, a constitution that could be a standard for valuation;

under- and over-exploited forests would have timber volumes and values greater and less than that standard.

Concurrently with acceptance of sustained yields, foresters rejected uneven-aged or selection working; in such forests the correct proportions of differently aged and sized trees are difficult to plan and accomplish and the fellings are difficult to control. Instead even-aged stands and even-aged forestry were adopted as the essence of good forestry. At the time it was first essential to organize and regulate fellings and yields, to prevent abuse and over-cutting where demand was keen. Forestry had to be re-organized and controlled and preferably also simplified in execution. Even-aged stands lent themselves to this type of re-organization, especially pure stands. Moreover, dense even-aged stands seemed to be copies of nature such as densely stocked virgin forests; therefore it was argued that they should produce maximum quantities of quality timber. The success achieved in the more or less even-aged stands that had been and were being grown under the Tire et Aire methods in France also supported even-aged forestry which can without difficulty be adapted to a precise, easily understood model to give sustained yields.

And so arose the universally accepted principle of sustained yields and the classic concept of the normal forest attached to and designed for even-aged stands. But if we accept that there must be a normal or ideal state for any forest whatever its silvicultural structure may be, it is clear that the normality will vary with the silvicultural system adopted.

The effect of silvicultural systems on normality

Silvicultural systems can be broadly classified in two main groups for the types given below:

A. *Even-aged systems*
   1. Clear cutting
   2. Strip
   3. Shelterwood
   4. Strip or group-shelterwood

B. *Uneven-aged systems*
   5. Group selection
   6. True (or single tree) selection.

The basic difference between the two groups is that in an even-aged system any small unit of area, such as an acre, contains trees of only one age although the forest as a whole contains all ages to maturity. In the uneven-aged forest, however, each small unit or area contains trees of all ages; in the single tree selection system the unit of area is very small, say one acre or even less, but in the group selection system it will be larger, perhaps five acres or even more, so that perhaps five mature trees occupy $\frac{1}{8}$ acre as a group.

In very intensive even-aged forestry, in order to take full advantage of differences in site and to cope with accidents, individual stands of one age may also be small, perhaps only $\frac{1}{2}$ or even $\frac{1}{4}$ acre. The forest then approaches and begins to merge into an uneven-aged one.

The basic difference between the two groups is whether separate ages and areas can be distinguished and delineated on the ground and on the map or not. In even-aged forestry they can be distinguished so that management can be based on area and age. If they cannot be distinguished, as in the single tree or true selection forest where the age of any tree is unknown and the area occupied by trees of any one age is incalculable, then management must be based on individual trees and their sizes—and *not* on age and area.

There must, therefore, be two main types of normal forest.

THE NORMAL EVEN-AGED FOREST

The classic normal forest, based on the clear cutting silvicultural system and sustained annual yields, postulates the presence of as many uniform aged stands, each having equi-productive areas, as there are years in the rotation. In other words there is a normal series of age gradations each of whose ages differs by one year up to the rotation age and each of whose yield capacities is equal. Each year the oldest gradation is felled and regenerated so that a sustained and equalized annual yield is secured. The area of each gradation, therefore, is A/R where R is the rotation in years and A is the area of the forest recorded in reduced or equi-productive units as explained in chapter III.

In addition each age gradation must be fully and ideally stocked, i.e. it must have the 'normal' volume which for each age is that stocking and volume deemed to be ideal for the thinning intensity adopted to achieve the objects of management. In other words the forest must have a normal growing stock.

Contemporaneous with the normal growing stock and normal series of age gradations there is also the need for all stands to be fully healthy and growing at the rate consistent with normal age and normal stocking. The forest must have a normal increment.

So there is a trinity of norms in the normal forest, namely the normal series of age gradations, the normal growing stock and the normal increment. Of those norms the age gradations depend on the ogranization of final fellings and the other two on silvicultural tending.[1] Consideration makes it clear that one and even two of the three elements may be normal in quantity over the forest as a whole even when that forest is far from normal. Thus, to take an extreme example, a forest might be composed of only one age gradation (of about half the rotation) and the volume of either the growing stock or the increment (or even of both) might equal the normal volume of a normal forest composed of all age gradations; the forest volume per acre of the one age gradation might equal the average volume per acre of all gradations of a normal forest. Similarly a normal series of age gradations might easily be understocked and consequently have a volume less than the normal volume so that the forest is abnormal. All three norms must be present for the forest to be normal as a whole.

It is not, however, at all necessary although it may be desirable for stands or gradations of one age to be in one compact area: they may be scattered among other gradations throughout the forest provided that their total area is correct. A ten-year-old gradation may be found in several stands in several places and may be next to a twenty-, forty-, or any other aged stand.

Practical modifications

Except when very short rotations are used, such as those for many coppice systems, it is seldom practical to distinguish between age differences of only one year. Five, ten or even more gradations may be grouped together to form an age-class.

---

[1] Thinnings, by their influence on the increment of individual trees, affect the length of rotation. The rotation controls the number of age gradations needed in the normal forest. To that extent the thinning intensity affects normality. But for a given rotation only final fellings and the consequent rate of regeneration controls the normal series of age gradations.

A forest subjected to a hundred-year rotation might then have, for example, ten age-classes, each class having stands aged one to ten, eleven to twenty years and so on to a hundred, each class occupying $\frac{1}{10}$ of the equi-productive area of the forest.

By dividing the rotation into an equivalent number of periods, for example ten ten-year periods, it might be specified that the oldest age class shall be felled and regenerated in ten years, preferably but not necessarily in equal yearly amounts. A degree of flexibility is then introduced which is consistent with the definition that a normal forest has that perfection which is practically attainable.

The concept of age-classes can be extended to forest areas. The forest can be divided into periodic blocks, perhaps but not essentially compact and self-contained, each of whose areas is preferably equi-productive. Thus, there might be five periodic blocks, each containing two ten-year age-classes for a rotation of a hundred years. Each block would be felled and regenerated in turn, during a twenty-year regeneration period, in as consistent and regular a manner as possible. An organization of that kind is not only simple but provides some flexibility which is both useful to meet market variations and essential to vary fellings that are designed for natural regeneration.

The normal growing stock of even-aged forests

Let us consider the volume of the growing stock of a normal, even-aged forest whose rotation is R years. Let us assume that the annual clear felling of the oldest age gradation is made after the end of the growing season and that it is regenerated before the beginning of the next season.

Then if $v_1, v_2, \ldots v_{r-1}, v_r$ are the volumes of each age gradation, the volume, V, of the whole forest can be calculated as below:

(i) After felling before the growing season
$$V = O + v_1 + v_2 + \ldots v_{r-1}$$

(ii) At end of growing season and before felling
$$V = v_1 + v_2 + \ldots v_{r-1} + v_r$$
Sum of the volumes
$$2V = 2(v_1 + v_2 + \ldots v_{r-1}) + v$$

(iii) In the middle of growing season and between fellings, the mean volume
$$V = v_1 + v_2 + \ldots v_{r-1} + \tfrac{1}{2}v_r$$

If yield tables are used to calculate the volumes of each age-gradation it must be remembered that yield tables invariably give volumes per acre for age-classes which are nearly always five or ten-year classes and not for age-gradations by yearly intervals. So each volume for the classes of the yield table in the mean volume per acre of five or ten age-gradations—e.g. for the five-year class the figure is the mean volume of stands aged three, four, five, six and seven years and for the ten-year class interval the figure at ten years is the mean volume of stands aged five to fourteen years.

So the yield table figures must be multiplied by the class interval. That means that the lowest age-gradations (i.e. gradations 1 and 2 for five-year classes and gradations 1–4 for ten-year classes) are ignored and an allowance has to be made for the rotation-age gradation. That allowance is $\frac{3}{5}$ of the volume of the age-class of rotation age since with a rotation of say a hundred years a five-year class would be the mean volume of gradations 98, 99, 100, 101 and 102 and a ten-year class the mean of 95–100, 101–104, of which we are concerned only with volumes to a hundred years.

So the normal volume of a series of age-gradations, can be calculated from the volumes per acre of age-classes given in a yield table by the formula:

$$V = \frac{C(v_1 + v_2 + \ldots \frac{3}{5} v_r)}{R}$$

Where V = the normal volume *per acre* of R age-gradations
    C = the class interval of five or ten years
    R = the rotation in years

$v_1, v_2 \ldots v_r$ = the volumes per acre of each yield table *age-class* to rotation age.

The calculation assumes that volumes increase regularly on either side of the yield table class figure. That assumption is approximately true but only for short class intervals and certainly not for intervals exceeding ten years. The fraction $\frac{3}{5} v_r$ assumes that the rotation corresponds exactly with the age of a yield table class, if it does not, e.g. a rotation of sixty-six years for five-year age-classes, the fraction must be adjusted accordingly to $\frac{4}{5} v_r$.

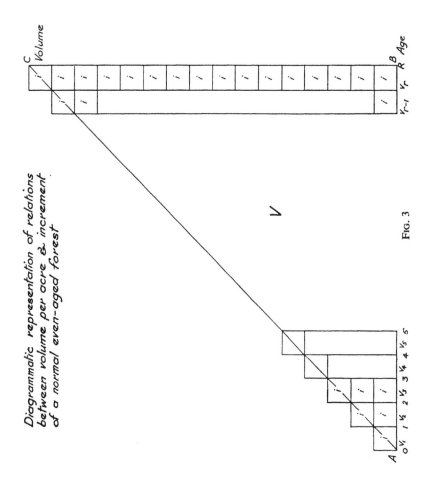

Relations between volume and increment

Let us consider Fig. 3 which represents diagrammatically the volumes per acre of a series of R age-gradations each one acre in extent of a normal forest whose rotation is R years.

Diagrammatic representation of relations between volume per acre & increment of a normal even-aged forest.

Fig. 3

Let $v_1, v_2 \ldots v_r =$ the volumes of each gradation.

Let it be assumed that the yearly volume increment, per acre, i, of each gradation is constant throughout the rotation. Thus, $v_1 = i$, $v_2 = 2i \ldots$ and $v_r = Ri$. The volume Ri also clearly equals the total yearly increment I, of the whole forest of R gradations.

Then in Fig. 3 each vertical column represents the volume of an age-gradation at the end of the growing season and the rectangle at the top of each column represents its increment, i, of the year.

The volume, V, of the whole forest, *in the middle of the growing season* when half the growth is completed, is therefore represented by the ABC whose hypotenuse AC bisects each rectangle of increment so that half of the increment is within the $\triangle$ ABC and half outside it.

so V = area of $\triangle$ ABC

$$= \frac{Ri \times R}{2} = \frac{I \times R}{2} \text{ or } I \times 0.5 \text{ R}$$

and $I = Ri = \dfrac{2V}{R}$ or $\dfrac{V}{0.5R}$

We must, however, remember that we have assumed that the MAI of the oldest stand $(v_r/R = i)$ applies constantly each year to each gradation. But in reality each gradation increases in volume by its own CAI and the CAI is *not* uniform each year but is slow in first youth, increases to a maximum and then progressively falls. So in fact the hypotenuse of the $\triangle$ ABC is not a straight line but is curvilinear and sigmoid in shape as shown in Fig. 4 so that the normal forest is really represented by the area under the curved line and not by the area under the straight line AB.

Now the straight line AB cuts the curved line so that towards the end of the rotation there is an area, *a*, above the straight line and towards the beginning of the rotation an area, *b*, below it. If those areas a and b are equal then the area below the curved line obviously equals the area of the $\triangle$ ABC which then truly expresses the volume of the normal forest.

But it is also clear that *a* and *b* can only be equal at one particular rotation, as shown in Fig. 4 by the $\triangle$ A'BC', at the rotation R–X for which *a* and *b* are shown shaded.

Therefore the two formulae, V = I $\times$ 0.5R and I = V/0.5R, are true only for one particular rotation. It is necessary to substitute a constant, c, for the 0.5. The two formulae then read:

$$V = I \times cR$$

$$I = \frac{V}{cR}$$

The constant, c, known as Flury's constant, will vary in quantity not only with rotation and consequent size of the normal growing stock but also with species and growth conditions. It must, therefore, be calculated for every rotation for each species and each quality or yield class.

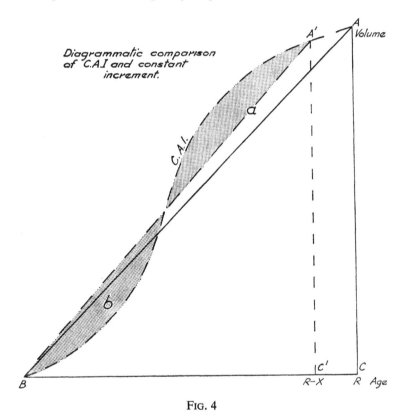

Diagrammatic comparison of C.A.I and constant increment.

FIG. 4

That can be done by using yield tables to calculate the true normal volume of each age-gradation and of the forest as a whole for various rotations and species and site qualities and relating the true volume with the calculated 'triangular' volume, thus:

$$c = \frac{v_1 + v_2 + \ldots v_{r/2}}{Ri \cdot R}$$

The value of c varies from about 0·4 to 0·6 for common rotations and European species, reaching 0·5 later on poor sites and for species that develop slowly in youth (whose CAI culminates late).

Thinnings. We have been tacitly assuming that no thinning takes place. That was to a great extent true in the early nineteenth century when the concept of the normal forest was first expounded and is still true for some short rotation crops such as coppice. But now, in Europe at least, thinnings are an important feature in the treatment of nearly all high forest stands. If thinnings are made the volume of the oldest age-gradation is less, by the amount of the thinnings that it has had, than the volume it would have had if left unthinned.[1] The MAI of the oldest age-gradation, calculated from the actual *standing volume*, does not, therefore, reflect the full increment that the stand has given; nor does the standing volume of the whole growing stock reflect the real increment that it is putting on at any one time.

Fig. 5 shows diagrammatically the effect of thinnings on Douglas fir (yield class 200) for a normal forest with a rotation of sixty years (British yield tables, 1966). The smaller △ ABC represents the volume of a normal forest calculated, as previously, from the standing MAI of the oldest age-gradation. (Again as previously, if the volume of each age-gradation were to be plotted the line AB would be sigmoid, not straight.) If, however, the volume of all thinnings previously made in the oldest age-gradations were to be added to its standing volume and its total MAI were to be calculated, the triangle becomes the larger △ A'BC. Then △ A'BA represents the volume of thinnings and the △ ABC represents the volume standing at the middle of the growing season in any one year.

The yearly increment of the forest is then represented by the area of the oldest age gradation along the line A'AC—final cut+thinnings to date, i.e. the *total* MAI of the oldest age-gradation equals the CAI of the whole forest when Flury's constant, c, equals 0·5.

If, therefore, the standing volume of a forest is measured and used to calculate its increment and consequent yield that should

---

[1] It has been shown that thinnings, within reasonable bounds of intensity, do not alter the total production of growth of a stand. Thus thinnings reduce in proportion to their intensity the final volume of the stand. The total yield is the same, only the time of harvesting that yield is affected.

be cut, we must remember that that yield applies only to the final cut and excludes thinnings. If, on the other hand, we measure the actual CAI of a forest, the thinnings are included. The two previous formulae, $I = V/cR$ and $V = I \cdot c \cdot R$, still hold good but the V applies to the larger A'BC, i.e. the standing volume *plus* thinnings to date. (See also chapter VIII, formulae methods of yield regulation.)

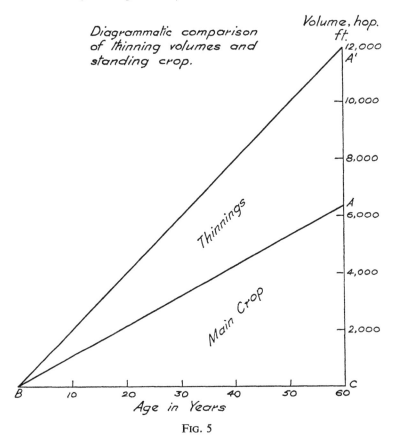

Diagrammatic comparison of thinning volumes and standing crop.

FIG. 5

THE NORMAL UNEVEN-AGED FOREST

In the entirely uneven-aged forest worked by the true selection system, trees of all ages (and sizes) are intermingled on every small unit of area such as 1 acre. The younger and smaller trees will occur in clusters, partly under older and larger trees and partly in gaps or openings of the upper canopy. The oldest and largest trees will be scattered everywhere in singles or perhaps couples. In such circumstances neither can the age of any tree be known nor the area of land occupied by any age-class.

Fellings, therefore, cannot be distinguished by either area or age; nor can thinnings (intermediate yields) be separated from final fellings and yields so that definite areas cannot be set apart each year for either final or intermediate yields. All fellings are, in fact, a continuous process of thinning a perpetual forest by selection of individual trees for cutting. Theoretically the fellings are scattered over the whole forest each year, but in practice the forest is divided into some five to ten sections, and fellings are confined to one section each year in turn.

Decisions on choice of trees for felling are based on size, health and increment, form and if necessary species. Large trees are felled when they reach the desirable, exploitable size or their increment falls below that acceptable. Other trees are removed on the general principles of thinnings to give the proper growing space to the better stems. All fellings are so made as to maintain or if necessary increase the irregularity of the stand and maintain or acquire the ideal proportions of large, medium and small trees.

Age and rotation have become meaningless and the only scheme in the spatial arrangement of the growing stock is the proper intermingling of different sized trees in their ideal proportions so that a regular sequence of maturing timber is secured—on the general assumption that, on average, size expresses age.[1] The numbers of trees of each size is, therefore, critical so that each year regular quantities of trees of each size are available for cutting.

Therefore, the normality of an uneven-aged, selection forest is judged by the numbers of trees in each size-class; it must have a normal series of *size*-gradations, instead of the normal series of age-gradations of the normal even-aged forest. In

---

[1] If rates of growth vary with size a smaller proportion of those sizes which are growing fast is needed to provide a regular sequence of maturing timber. There is no untoward difficulty in assessing the *average* rate of growth of each size class. But individual trees in a class may acquire a lower performance and yet appear healthy. They may then be left unthinned and a strain or provenance of poor performance may thereby be encouraged.

addition it must also have the normal volume and normal increment as well as the amount of irregularity per acre (in small groups or in single mature trees) that is deemed to be the most satisfactory.

Although it is obvious that a normal uneven-aged forest must contain more small trees than big ones per acre, it is difficult to devise a simple model, such as the triangle of normality for even-aged forest, to represent either the numbers or volumes of trees in the several size-classes. Nor are there yield tables to show what the normal numbers of trees should be in each size-class.

There is, therefore, uncertainty of what the constitution of a normal uneven-aged forest should be and one is led to Biolley's concept of the normal growing stock as that which produces permanently the most valuable increment. That ideal state for any particular forest can only be found by long experience of working it, guided by knowledge of other similar forests thought to be normal. By keeping exact comparable records, frequently and periodically compiled, of the standing growing stock and its distribution in each size class and of similar figures of what is felled, the progress of increment and yields can be watched. Having that knowledge the growing stock can be moulded to that which is the most productive of valuable increment. That is the method, known as the Method of Control used in Switzerland, where selection working is perhaps more successful than anywhere else and more widely practised.

But yield tables for normal even-aged growing stocks can give figures to illustrate what the constitution of an uneven-aged forest may resemble.

If one plots on graph paper the stem size-frequency curve for any one age-gradation one will get a Gaussian curve, as an inverted U, which is more or less normal in form, peaking at the mean breast height (B.H.) size. But if one takes from a yield table the numbers of stems for the mean B.H. size of each of all the age-classes up to rotation age and plots the numbers against size on squared paper, the curve will be a reversed J curve. The curve represents the numbers of trees per acre by sizes throughout a normal sustained unit of R acres.

Fig. 6 shows such a curve for Norway spruce, yield class 200, compiled from the British Yield Tables. In that figure the size-frequencies were taken from the mean quarter girth at breast

height (QGBH) of each of the sixteen five-year age-classes up to
a rotation of eighty years. Those frequencies were then divided
by sixteen to give figures for all sixteen age-classes reduced so

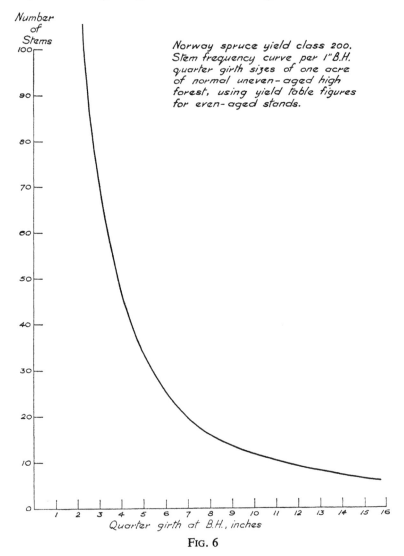

FIG. 6

that they occupy together only one acre instead of 80 acres—as
they would in 1 acre of normal uneven-aged forest. The size-
frequency curve of Fig. 6 thus represents a normal sustained

yield of even-aged Norway spruce on one acre with a rotation of eighty years comparable with one acre of normal selection with an exploitable size of 15¾ inches QGBH and an exploitable age of eighty years.

We can see, therefore, that in both uneven and even-aged forest there is a normal series of size-gradations, depicted by a reversed J curve. But in even-aged forest the curve applies to the *whole* felling series or forest; in the uneven-aged forest it applies to *each small unit of area*, such as two acres, throughout the forest. Consequently in the normal selection forest each small unit of area becomes a sustained yield unit whereas in even-aged forests the sustained yield unit is the whole felling series which is composed of a whole normal series of age-gradations.

Table 6 summarizes the frequencies, read off from the curve of Fig. 6, for 1 inch QGBH classes. Column 3 of the table shows the coefficients of diminution by which the frequency of one size is reduced to that of the next larger size.

Table 6

SIZE-FREQUENCIES FOR NORWAY SPRUCE (200 YIELD CLASS) ON ONE ACRE WITH 16 AGE CLASSES TO 80 YEARS

Size Class Q.G.B.H. inches	No. of Stems	Coefficient of Diminution
1	2	3
3	74	1·54
4	48	1·41
5	34	1·36
6	25	1·25
7	20	1·25
8	16	1·22
9	13·5	1·15
10	11·75	1·15
11	10·25	1·14
12	9	1·12
13	8	1·11
14	7·25	1·11
15	6·5	1·12
16	5·75	

It is noticeable that the coefficient of dimunition is not constant but falls steadily from 1·54 to 1·11 (at the 13 to 14 inch class) and is then steady.

4*

It was suggested by F. de Liocourt in 1898 that the number of stems in a selection forest lessens from one size-class to the next larger class in a geometrical progression, so that the diminution is constant. If that were to be true the series could be expressed as a, ad^{-1}, ad^{-2}, ... $ad^{-(n-1)}$, where a is the number of stems in the smallest class, n is the number of classes and d is the coefficient of diminution.

If that theory is true and if the numbers of trees in any size-class (e.g. the smallest one) and the coefficient of diminution are known, the whole normal frequency can be calculated. Now the number of trees needed in the smallest size-class depends on two things—the time taken for trees to grow out of the class into the next (time of passage) and the number of trees (of all sizes) that can be felled each year without depleting the growing stock—i.e. when fellings equal the increment.

Thus if six trees per acre can be felled yearly and the time of passage is ten years, there must be sixty trees in the smallest class to replace the trees being felled. (With a ten-year time of passage $\frac{1}{10}$ of the trees in a class move to the next higher class yearly; $\frac{1}{10}$ of 60 = 6.)

The time of passage (rate of growth) can be measured but the number of trees (increment divided by average felled tree volume) that can be felled without depleting the growing stock is obtainable only by experience or comparison with another forest.

If we can then decide (from experience and comparison with other forests) what size of tree it is desired to grow so that there is everywhere one per acre, we can state that:

$$ad^{-(n-1)} = 1$$
$$a = d^{(n-1)}$$
$$\text{and } \log a \quad (n-1) \log d$$
$$\log d = \frac{\log a}{n-1}$$

Having already calculated a and having decided n by fixing the large size class at one per acre, d is calculable and also the whole frequency.

It is clear from the above that the calculation of the normality of a selection forest is fraught with difficulties and uncertainty even if de Liocourt's theory is accepted. The time of passage is the only measurable factor and even that varies with the density of the stand and that density varies with the indefinable growing stock.

So the constitution of the best selection growing stock of any particular forest is only truly known as the result of experience. Continuous comparable periodic measurements of numbers and volumes of trees, both of those felled and those left standing, classified for each size-class and each compartment of the forest, are essential over many years as is done in the method of control.

Then the increment (I) of the forest as a whole, of each compartment and each size class is easily assessable:

$$I = V_2 - V_1 + F_v$$

Where V_1 is the volume at the first measurement, V_2 is the volume at the second measurement and F_v is the volume felled.

Provided that the periodic measurements are entirely comparable a full knowledge of the growing stock, increment and yield and their progress is obtained, size-class by size-class, compartment by compartment and for the forest as a whole. The result of treatment (the fellings) can then be assessed and the fellings can be adapted to develop the correct proportions of size-classes to give the greatest valuable increment for the forest.

Timber prices normally increase with size of tree to an optimum size so that the most valuable yield is attained when the size of the mean tree felled is a maximum. It is desirable to have as many large trees as possible provided that there is constant recruitment from the smaller sizes. A critical factor to watch, therefore, is the volume of the smaller size-classes. If that volume remains constant, regeneration and recruitment are satisfactory. If the volume tends to shrink regeneration and recruitment are failing; more space must then be given to regeneration by reducing the volume of the growing stock by felling more large trees.

For any particular forest, therefore, there will be a maximum ideal volume of growing stock per acre as well as the ideal proportion of size-classes. The proportion of size-classes is reflected in the volume of the mean tree. Thus, for any one forest there will be an ideal growing stock volume and an ideal mean tree volume.

For the forest of Couvet in Switzerland it appears that the ideal growing stock volume is about 360 silves[1] (about 360 cubic metres) per hectare in which the mean tree has a volume of 1·5 silves. For the forest of Les Joux M de Coulon (1962) suggests that the ideal total and mean tree volumes should not exceed 400 cubic metres per hectare and 1·2–1·3 cubic metres respectively. If one silve is taken to equal 1 cubic metre the volumes per acre at Couvet and Les Joux are respectively about 4040 and 4490 hoppus feet. The annual increment at Couvet is about 9·7 silves per hectare or 108 hoppus feet per acre.

DISCUSSION

1. *Volumes of normal growing stocks.* As has been shown above the normal volume of the silver fir—Norway spruce—beech selection forests of Couvet in Switzerland, grown to produce large trees with an exploitable age up to about 150 years, may be 360 silves per hectare or, say, 4000 hoppus feet per acre.

Biolley (1920) for the Neuchatel Valley of Switzerland gave volumes in cubic metres per hectare of 301–400 for rich forests, 201–200 for average forests and 107–200 for poor quality forests. In general therefore the normal volume of selection forest in Switzerland may vary between about 1250 and 4500 hoppus feet per acre, according to quality of the site and exploitable age and size.

For even-aged coniferous forest Table 7 shows the normal volume per acre for pure Norway spruce, Douglas fir and Scots pine given to a rotation of eighty years. The figures are taken from the British yield tables of 1966 (Bradley *et alii*, 1966) adapted to ten-year age-classes (five-year classes in the tables). The volumes given would naturally be greater for longer

---

[1] A silve is the volume unit for standing trees whose volumes are estimated from breast height measurements and volume tables. Nominally a silve is 1 cubic metre but being an estimated volume the true volume of the tree, measurable in cubic metres after felling, may differ from the silve volume. If the same volume tables are used continuously, however, silve volumes of successive measurements are always comparable. (If the tables are +10 per cent wrong, the comparison of repeated volumes each +10 per cent wrong is true.) The relation between silves and the cubic metres can be obtained in any forest empirically.

Table 7

VOLUMES OF GROWING STOCK IN A NORMAL SERIES OF TEN YEAR AGE CLASSES TO A ROTATION OF EIGHTY YEARS ON 80 ACRES

Species	Yield Class	Volumes of each 10-year age class							Total Volume of Series	Average Volume per acre	MAI per acre at 80 years	Age of Maxm. MAI
		11–20	21–30	31–40	41–50	51–60	61–70	71–80				
Norway Spruce	240	6,313	20,837	33,925	50,350	63,225	72,838	82,200	329,688	4,121	233	65
	200	4,533	14,382	28,350	42,450	54,050	64,688	72,287	280,740	3,509	197	65–70
	160	1,800	11,238	22,187	34,063	44,150	52,587	60,350	226,375	2,830	159	70–75
Douglas Fir	240	9,437	24,800	38,875	51,975	63,700	68,183	81,345	338,315	4,229	221	50–55
	200	7,288	19,187	32,988	44,337	54,338	63,425	70,375	291,938	3,649	186	55
	160	5,538	15,048	26,420	36,482	44,850	52,388	58,087	238,813	2,985	150	55–60
Scots Pine	160	6,300	14,025	23,325	33,400	42,538	50,775	57,825	228,188	2,852	157	65–70
	120	3,750	9,000	16,438	24,862	32,875	39,500	43,150	170,475	2,131	119	70–75
	80	—	5,000	9,938	15,662	21,875	27,500	34,625	114,600	1,432	80	80

rotations so they are not comparable in any way with the volumes given above for selection forest in Switzerland where species are different and the exploitable age is up to 150 years.

If maximum volume yields are wanted the rotations in Britain would be, for Douglas fir fifty to sixty years, for Norway spruce sixty-five to seventy-five years and for Scots pine sixty-five to eighty years, depending on the site quality and yield class. Such rotations would give smaller average volumes for the whole forest than those for an eighty-year rotation as shown below:

Species	Yield class	Rotation (Years)	Average vol. of whole series (hop. ft./acre)
Douglas fir	240	50	2,502
	160	60	2,139
Norway spruce	200	70	2,978
Scots pine	120	70	1,819

2. *Values.* In Britain, after allowing 10 per cent for under-stocking, one might reasonably expect an average volume of 2000 hoppus feet per acre in an average normal forest evolved for maximum volume production with various coniferous species and yield class. If the timber is worth 1s 6d per hoppus foot standing the value of 2000 hoppus feet per acre is £150 or £15,000 per 100 acres. Values of £20,000 per 100 acres for some species and some sites might easily be obtained and for long rotations volumes of 4000 hoppus feet per acre and values of £30,000 or more per 100 acres might be expected.

Those tentative figures of value are given only to emphasize the importance of the capital tied to the growing stock, such as £150,000 in a normal coniferous forest only 1000 acres in extent. The supreme importance of the growing stock becomes clear and the need for its effective and efficient manipulation.

3. *Types of normality.* That supreme importance of the growing stock in all planning for efficient management demands an ideal type of growing stock at which to aim. As has been pointed out before, the classic concept of the ideal or normal forest was based on sustained yields from each forest or even from each of several parts of one forest. The principle of sustained yield is basic to the classic concept of normality. If that principle is no longer paramount for a production forest

(excluding for the moment a 'service' forest managed for recreation, protection and the like), the need for an ideal norm remains but has to be re-defined. What is it?

Without the need for a sustained yield the need for a normal series of age-gradations disappears; some other basic object of management must take the place of sustention. In production forests that new basic object will probably be sustained production or growth of the required product. Emphasis will now be placed on silvicultural tending of the growing stock so that each site is fully stocked with trees of the species (within the bounds of the product needed) best suited to that site. These stands will be so thinned and tended to give the maximum increment to the size or age required by the objects of management; they would be removed and regenerated as soon as they fail, by reason of age and size, disease or accident, to produce increment at an adequate level.

In such a normal forest, by forsaking attention to the sequence of maturing timber greater attention can be paid to the productive capacity of stands or even individual trees in selection forest. No longer is sacrifice from felling under- or overmature trees necessary to attain normality but, rather, normality is promoted by constant removal of any stand that is behaving below standard. The consequent sacrifice of having a fluctuating yield is no longer important.

Yields might, indeed, fluctuate markedly particularly when it is found that another species will perform better than the one currently used in various sites. But although yields in one forest unit might fluctuate, perhaps violently, the continued yield from several forests in a region would, at least after some period of management, tend to be sustained and uniform. The fluctuation in yield from individual forests would cancel each other, especially if there is a variety of sites species and rotations.

In fact the use of rotation in its strict sense of the planned number of years between formation and harvesting of a stand would disappear to be replaced by an infinite number of removal ages for individual stands. The rotation would become the average of the removal ages and be comparatively unimportant in planning except for forecasts of future yields.

In such a normal forest regulation of the yield (see chapter VIII) becomes less important except for controlling the intensity of thinnings. Final fellings would depend not on control of sustention as in the classic normal forest but on the performance of individual stands. But the need to know what the outline will be for planning sales and general forest work naturally remains. Yield regulation will be replaced by forecasts of future outturn, both from thinnings and final fellings which depend on the current performance of, mainly, older stands. Judgement of the state of performance depends on the objects of management, such as maximum volume production, maximum net discounted revenue, and particular class of timber and so on.

In some forests, such as in Victoria (Australia), the paramount object of management may be the maximum sustainable yield (Orr, 1966). That means that the yield should rise continuously to some sustained maximum but should never fluctuate below some recognized minimum so that employment is ensured and wood industries suffer no lack of raw material from the local forest unit. In effect that means that the normal forest is the classic ideal with a sustained yield but its attainment is delayed by the need to vary fellings to prevent downward fluctuations in any year; the yield must be steady or rise steadily. Both thinnings and final fellings may be either delayed or advanced for that purpose, so that yield regulation based on forecasted states of the growing stock is a basic factor.

4. *Service forests.* Normality in 'service' forests is also desirable to promote effective and efficient planning and treatment, but there may be great variety in normality dependent not only on the purpose of management but the locality, its siting and topography. In many service forests, e.g. those for protection of the land in mountains and frequently those managed for their beauty, fellings will be as few as possible, single tree-selective in type with a high exploitable age. Shelter against wind by belts of forest may also require selective working, with a heavy undergrowth, so that the shelter is never interrupted by local clear cutting. In others the desirable uneven-aged structure may be influenced by the need for open glades or spaces and variety of species to give different tints and shades of colour. In forests managed for beauty and recreation sudden changes are unwelcome and again postulate a perpetual forest worked under gentle selection fellings, by individual trees or small groups. In other recreation forests, managed perhaps in relation to

lakes and streams, camping or even hotel facilities may become an essential part of the ideal or normal service forest.

But whatever the objects of the service forest may be an assessment of its most desirable form should be made together with the consequent treatment it will need, even for an ecological preserve kept for study of its fauna and flora. Forests always change at least locally even if the mean state of a large virgin forest remains constant; it develops, waxes and wanes in a series of successions. To bring a forest to a wanted structure requires treatment; to keep it in that state, whether it is entirely used or not, requires treatment of some kind without which it will tend to lose its ability to supply what is wanted. The definition of what is wanted, the ideal or normal forest, remains necessary so that it can be attained and kept.

Bibliography

Anon. (1953) 'British Commonwealth Forest Terminology', part I. Empire Forestry Association, London

Biolley, H. E. (1920) *The Planning of Managed Forests*, translated by M. L. Anderson (1954). The Scrivener Press, Oxford

Bradley, R. T., Christie, J. M., Johnston, D. R. (1966) *Forest Management Tables*, Forestry Commission Booklet No 16. HMSO

Coulon, M. de (1962) 'Structure et Evolution de Peuplements Jardinés des Joux', *Journal Forestier Suisse*, 113 (10)

Orr, R. G. (1966) 'An Evaluation of Methods of Yield Estimation for use in Conifer Plantations in Victoria.' Unpublished diploma thesis at the Commonwealth Forestry Institute, University of Oxford

Chapter VI

THE ORGANIZATION
OF FORESTS

In comparison with other uses of land forests are large, sometimes very large covering hundreds of square miles, often
scattered in many pieces so that one forest may be used for a
variety of purposes and worked in several different ways on
separate time-tables of work and with separate units of man-
power. Forests must, therefore, be divided into several parts for
convenience of description and record, location of events and
for general administration as well as for the organization of
work programmes and methods of silvicultural management.

Some subdivisions of a forest are internationally used and
recognized but in others there are differences in nomenclature
or definition and use which confuse understanding of literature.
Some definition and discussion of the subdivision of forests and
the terms used is necessary.

A. PRIMARY TERRITORIAL UNITS

1. *Woods, blocks and enclosures*

These more or less synonymous terms refer to wooded areas,
often bounded by natural features, which have well-known
local names. They may have resulted from legal separation by
enclosure from surrounding land for preservation or distinction
of ownership. Except for any legal properties that may be
possessed by such areas they have no managerial significance
except for their convenience for location and territorial sub-
division of a large forest, a sub-division which may often be
used for allotment of administrative charges.

This type of block should not be confused with Periodic or
Working Blocks (see below).

2. *The compartment*

The compartment is the smallest permanent sub-division of
a forest. It has been defined (B.C.F.T., 1953) as 'a territorial unit

of a forest permanently defined for purposes of administration, description and record (preferably designated by Arabic numerals, 1, 2, 3, etc.)'. To that definition may be added the purpose of location.

Accordingly we may note that the compartment is not a unit of treatment although it may be used for that purpose in some countries, e.g. New Zealand, or for some methods of working as for the shelterwood compartment silvicultural system in France.

Being a permanent unit the compartment should be clearly demarcated on the ground and its boundaries should follow natural features such as ridges and water courses, or definite artificial features such as roads, rides, main ditches, banks or other surveyed lines. The shape of compartments should be compact and in flat, featureless country where artificial delineation is necessary compartments may usefully be rectangular or square and have a convenient area such as 25, 50 or 100 acres or simple fractions of a square mile such as 32, 64, 128 or 320 acres with boundaries made to serve as roads, rides and fire-lines.

The size of a compartment depends primarily on the intensity of working and also on topographic or site variations which may affect the type of management and its intensity. In large, extensively worked forests, such as many in the tropics, compartments may be 500 or even 1000 acres and more in extent. In the intensively worked selection forests of Switzerland where the compartment may also be a unit of treatment, the area may be only 15 to 20 acres. In England compartments average some 25 to 30 acres as a rule. In France the combined factors of ecological conditions, structure of the growing stock and conditions of exploitation and sale control the disposal of compartments which are managerial units for allotment of coupes and other work. In high production forests where the intensity of work is high the maximum size that is recommended in France is 25 acres (Anon. 1965).

A compartment is not identified with the composition or state of the growing stock or with site variations. The former will clearly change with treatment and natural development— in specific composition, age or stand structure—so that the growing stock cannot be a factor in distinguishing a permanent forest subdivision. But efforts have been made, as in Uganda

(Eggeling, 1947), to co-ordinate compartments with site-types and vegetation associations but proved impractical. Site-types are liable to be too variable in distribution, size and shape to suit a permanent unit which is also compact and has a substantial size. Moreover different sites require different treatments and growing stock composition so that they are more suited to be units of treatment such as the sub-compartment (see below).

It is clear, therefore, that different parts of a compartment may be subjected not only to different methods of treatment but also different objects of management.

3. *The sub-compartment*

The sub-compartment is a unit of treatment and is defined in the BCFT (1953) as 'a subdivision of a compartment, generally of a temporary nature, differentiated for special description and treatment (preferably designated by small letters, a, b, c, etc.)'.

The sub-compartment is identified with the growing stock and consequently with the site which naturally affects the type of growing stock and its development and treatment. Variety of site and treatment of the growing stock results in the formation of different stands which can be defined as collections of trees or other vegetation which are sufficiently homogeneous in specific composition, structure, age and rate of development or health to differentiate them from each other for purposes of description or treatment.

Accordingly the sub-compartment is usually synonymous with the stand at least in intensive forestry and has a silvicultural and utilization function while the compartment has a managerial, administrative function. Being temporary, sub-compartments do not need demarcation on the ground although they should be delineated in maps and their areas calculated and recorded.

The shape of a sub-compartment will be more irregular and its size, dependent on the intensity and detail of working, much smaller than the containing compartment—unless the compartment has so uniform a site and growing stock that no subdivision of it is necessary. In extensive working small, heterogeneous stands would be ignored and included with the surrounding stand whereas in intensive working they would be separated as distinct stands. Even in intensive working it is unusual and impractical to distinguish stands which are less than 1 acre or $\frac{1}{2}$ hectare in extent and they are usually larger. Unless the map scale used is very large it is impractical or impossible to map areas smaller than about an acre. In France it is considered undesirable to have more than three sub-compartments in a compartment and the minimum size should be about one quarter of the compartment (Anon. 1965).

Discussion on compartments and sub-compartments

In effect the compartment is a managerial microcosm or sample of the whole forest in the same way as a single shop is a sample of a large retail trading firm. To enable an enterprise to achieve the best results each unit must contribute its own best so that a knowledge of the performance of each unit is essential for managerial efficiency. Each unit can have its own peculiarities; thus two compartments may have the same growth potential but the one may be accessible the other inaccessible to transport. Records of the results of working the two compartments should enable analyses to be made to show whether it is profitable to improve communications to the one compartment or not. But to have a practical record value of that kind compartments should not be too small.

A basic factor in successful forestry is the attainment of the utmost production of what is wanted from each site. Treatment of stands (or sub-compartments) always aims at that—whether the product is recreation, profit or volume of timber. Fundamentally the forester always wants to know how to get that maximum production at least cost, and the performance of each stand is the witness of the degree of his success. So records of the history and performance of each stand or sub-compartment are essential.

Research in general should investigate and collect figures of volumes and rates of growth for different species and quality or yield classes under different thinning regimes and publish results in general yield, stand and thinning tables. But to use the tables locally each stand must be related to them—by age, quality, density—since stands may develop differently from the average progress shown by the general yield tables. Full knowledge of the history of each stand or sub-compartment is necessary to assess results of treatment and to guide current treatment.

Accidents such as fire, variations in development of a stand owing to site variations or local disease will break up stands into smaller fragments while at regeneration there may be either more fragmentation or some unifying enlargment. There is consequently an ebb and flow of change in stand units so that records will not apply to a constant unit of area. But the sum of the sub-compartment or stand records in any compartment does apply to a constant area, so that the summarized compartment and managerial records are comparable from time to time. Similarly stand or sub-compartment records are used for incorporation in the records of larger units of silvicultural organization such as the working circle and felling series which do not necessarily conform with compartments (see below).

For compartments to accomplish their full administrative and managerial function it has been suggested that they should not be too small (Osmaston, 1957). They might be enlarged by grouping three to five or more compartments together as parcels (of 100–200 acres in Britain) to form compact seasonal work areas based on configuration of the land and communications. They would be similar to but smaller than the 'working block' (used by the British Forestry Commission) and the cutting section (see below).

An advantage of such a parcel or group of compartments whuld be that the average size of stands would also be larger since stands would not be fragmented to conform with the boundaries of small compartments but only with those of parcels. Records would in consequence be easier and less laborious to compile and have a greater managerial value now that mechanization requires more concentration of seasonal work. Records would be filed with the parcel or group.

Compartments which are too small defeat the object of administrative convenience as they become entirely separated from organization of work. If they are too large they lose their values of location and permanent units of record but if they are too small they complicate unnecessarily the keeping of their records. Provided that compartments are adequately divided into suitable sub-compartments by stands, and provided that the stands are the true unit of working, much larger compartments, whereby there are only ten to fifteen compartments in a sub-manager's area (the Forester or Ranger's charge), seem advisable. Exceptionally, however, small compartments would still be necessary whenever they are used as silvicultural units of treatment such as in the Selection or Group-Selection Systems.

Finally it is worth stressing that compartments and their boundaries should not be changed. If they are changed the value of their past history is largely lost. If they must be changed—by grouping into parcels or by sub-division of excessively large ones—then whenever possible each change should involve the same unit of land so that a parcel contains an exact whole number of the old compartments or a large compartment is split into an exact number of whole new ones. The records of the original compartation can then be applied and compared with the new.

B. UNITS OF SILVICULTURAL ORGANIZATION

1. The working circle

The B.C.F.T. (1953) defines a working circle as 'an area (forming the whole or part of a working plan area) organized with a particular object and under one silvicultural system and one set of working plan prescriptions. In certain circumstances working circles may overlap.'

The need for this sub-division of forests that contain distinct forest types with distinct objects of management is clearly necessary for organization of work. For example a forest may contain hardwoods which it is desirable to work partly as coppice for supply of small wood products to local agriculturists and partly as high forest to yield saw-timber for more distant markets; each part would need a very different silvicultural and managerial organization so that their separation into two working circles has clear advantages. If the same forest also contains coniferous forest those areas would require yet another silvicultural treatment and a third working circle while still other areas might be set apart for recreation and amenity so that four working circles would be desirable in that forest. In each working circle there would be a concentration of purpose and method. Similarly a coniferous forest might be worked with an irregular, uneven-aged silvicultural system in the hilly tracts and with a uniform, even-aged system in the plains and lowlands. Owing to the fundamental differences in methods of applying the two silvicultural systems two working circles would be necessary.

It is clear that allotment to working circles depends on their position and factors of site which together affect the type of product wanted, the specific composition possible and the type of treatment. Units composing a working circle will, therefore, be sub-compartments or stands and *not* compartments. Consequently a compartment may contain parts of one, two or more working circles and a working circle may be scattered in many pieces over a forest so that it is not necessarily compact although compactness may be managerially desirable but unattainable owing to site variations.

Rigid application of the principles of the formation of working circles would increase their number and dispersion. Thus it might be argued that differences in methods of regeneration, rotation and type of product (such as telegraph poles and saw-timber) might require stands of Norway spruce, Douglas fir and Scots pine to be placed in three different working circles even if each species is worked under a similar even-aged system of silviculture. That is an unnecessary complication. Provided that each species is worked under basically the same silvicultural system, such as an even-aged system of high forest, and is subject to basically the same object of management, such as highest net income, or best financial yield or highest volume production, so that treatment of each species is similar but variable in degree, the several species should be included in one working circle. The main complication would be variety of rotations and consequent difficulty in making the working circle a sustained yield unit which has been the classic attribute of working circles (see Schlich (1925), who defined a working circle as a separate series of age-classes). But that complication also applies to a working circle composed of only one species growing on a variety of sites and having different rates of growth and, therefore, different rotations.

Schlich, however, advocated rigidity in allotment of areas to working circles in the interests of uniform sustained yields. Even for one species worked under one silvicultural system, he advocated two or more working circles if more than one rotation was to be used on the grounds that either the yield would be uneven, or if the same quantity was to be cut yearly the different rotations would merge into one. But Schlich qualified excessive multiplication of working circles by stating that 'moderate differences of conditions, especially in the rotation, should not induce the forester to introduce separate working circles'. Today inflexibility is quite unjustified. As explained in chapter III, the mobility of both man-power and distribution of products makes the need for local equalized annual yields no longer essential (except in undeveloped areas where the community depends on local forests), even if they are desirable. By using an average rotation for the general organization of fellings in a working circle and by having variable removal ages for individual stands, sustained production will be eventually achieved and periodic yields should not be too uneven (see also chapters IV and VIII).

If the more flexible approach is accepted it might be possible to extend the working circle still further to include even wider rotation limits to include hardwoods with conifers in one working circle provided that the silvicultural systems used are compatible so that thinning and utilization programmes can be co-ordinated in one set of prescriptions. An example of that having been done is in the Forests of Dean and Highmeadow Working plan (Anon. 1960).

Among other points to notice is that working circles should remain unchanged for as long as possible. They form a basic part of the silvicultural organization and each circle implies the use of one silvicultural system which should be an element of long-term planning. To convert one system to another is seldom easy and always involves some sacrifice; changes should be made as seldom as possible. If it is decided that it is essential to convert an area, the area should be transferred at once to the new working circle and not later even if the actual destined conversion of the stands involved is unlikely to be made for some years. For example stands of middle-aged oak may be chosen for conversion and allotment to the conifer working circle. The allotment to the conifer working circle should be made at once so that the influence of the oak stands on the conifer working circle as a whole—such as the sequence of maturing timber—can be constantly considered. To keep the oak stands temporarily in the oak working circle until they are mature confuses planning in each circle.

Overlapping working circles may be necessary if a forest contains a mixture of species one of which requires silvicultural treatment intrinsically different from that needed by the others provided that the two treatments do not unduly hamper each other. A good example from the tropics is hardwood forest containing bamboos. The latter are usually worked by selective cutting on a short felling cycle of three or four years; the trees may be worked under an even-aged system or a selection type of felling with a long cycle of fifteen to thirty years. Although the two methods of working the trees and bamboos are entirely different they do not greatly interfere with each other; the bamboo working circle then overlaps one or more tree working circles.

In the U.S.A. the term working circle is defined by Meyer, H. A. *et alii* (1961) as 'an area of forest requiring a separate working plan and for which the objective will be a sustained yield'. It is, therefore, akin to the term working plan area used in the British Commonwealth. In the U.S.A. the synonym for the working circle is working group which the above authors describe as 'a subdivision of a working circle on the basis of similar types. It is the silvicultural subdivision of a working circle.' The authors also state that in British Columbia the term working circle (in the sense used in the U.S.A.) is being dropped in favour of 'a sustained yield unit'.

In France the working circle is not used; its place is taken by the Series which may be defined as 'a forest or part of a forest allotted to the same kind of treatment and constitutes a distinct unit of management' (Anon. 1965). Its size, however, is usually limited to 1000–2000 hectares so that a large forest may contain several series although each may have the same kind of treatment. The French series, therefore is more akin to the felling series (see below) of the British Commonwealth terminology.

2. *The felling series*

A working circle may be too large and unwieldy, too scattered over the working plan area which may itself be far-flung, for convenience of administration and for a balanced distribution of both the yield and work in general. The working circle is then divided into two or more units, which are not necessarily comparable in size, each unit being complete in itself. These units are felling series whose definition in the B.C.F.T. is 'an area of forest delimited for management purposes and forming the whole or part of a working circle. Its objects are (1) a distribution of felling and regeneration to suit local conditions and (2) to maintain or create a suitable distribution of age classes.

The yield is calculated separately for each series which should have an independent representation of age-classes.'

It is evident that the concept of the felling series implies that the general policy and prescriptions of work, such as silvicultural system, for the containing working circle apply to each of its felling series but the application of the policy will vary with conditions in each series. Moreover, each felling series should be a sustained yield unit so that its yield is regulated independently of its fellows; consequently each series has its own thinning, felling and regeneration plan—in location, quantities and times.

Multiple felling series distribute all work more equally over the whole working circle in smaller thinning and felling coupes. That more scattered distribution is particularly marked when the method of silviculture involves the use of compact annual coupes which move in a definite direction. Examples are many. Thus selective working of a tropical forest on a twenty-year cycle usually implies the division of the forest into twenty more or less compact and large sections which are worked in turn; fellings in the tenth year may then occur 20 miles away from those made in the first year in section No. 1 where the felling is not repeated for twenty years. Similarly, on a smaller scale, in coppice working whole coupes are clear-cut yearly in sequence. If the coppice working circle is large and undivided into felling series, each coupe will be large and in one place; then if, as often happens, the working circle is distributed round the margins of a large forest, coupe No. 1 may be several miles away from coupes 7, 10 or 15, so that one locality is starved of work and yield for several years and another is temporarily overloaded.

In such methods of working, particularly where communications are poor, there are obvious advantages in the formation of several felling series. By allotting a separate series to each sub-managerial or executive charge, such as that of a forester or ranger, work is divided equally among the staff and labour force. If the charges and series also coincide with divisions of catchment areas, such as on either side of a ridge of hills or of a wide river and so also coincide with communications to different centres of demand, sales and supplies to separate markets are also benefited. Felling series also offer advantages in organizing leases to contractors for felling and extraction of timber; lease of work in each series can be negotiated separately, thereby facilitating both control and competition. Additionally multiple felling series provide assured local employment. Silviculturally there are also many advantages in small annual coupes.

The advantages and basis of formation of felling series are therefore mainly administrative and for control and distribution of sales, produce and employment; the silvicultural advantages of small coupes could be obtained in other ways. The main advantages, therefore, acrue largely in forests managed for sustained or regular yields which become less important with improved communications and transport. The latter also encourage and enable the practice of an intensive silviculture which is based on individual site and stand treatment. Such treatment will result both in scattering fellings and in reducing the size of individual coupes. Nevertheless it will still be administratively useful to form separate felling series for each executive charge, but for each series to be strictly a sustained yield unit becomes unnecessary. Variation in the performance of small, separate stands, worked without any predetermined regulation of felling progress, will in itself tend to make periodic yields uniform.

3. *The cutting section*

When fellings proceed in a definite direction through a felling series successive yearly coupes adjoin one another. Then, under clear-felling, there may be undesirable exposure of young growth to sun and wind while insect damage may be encouraged. To mitigate disadvantages of contiguous coupes and yet retain progress in a definite and orderly direction, a felling series can be divided into several cutting sections in which annual fellings are made in turn. Thus with five cutting sections fellings would be made every fifth year in any one section so there would be a five-year growth interval between contiguous coupes. If the number of cutting sections tallies with the thinning cycle, work each year can be concentrated in one section and repairs to roads, rides and drains can be similarly co-ordinated by sections.

4. *The working block*

To organize work efficiently in each beat, or subdivision of a forester's or sub-manager's charge, the State forest service in Britain concentrate operations such as brashing, thinning, pruning, maintenance of drains and fences, in a group of compartments each year. Such a group is known as a working block. The number of working blocks in a beat will depend on the length of cycle prescribed for the operations concerned; consequently there are usually three or four working blocks in a beat.

The working block is, therefore, very similar to the cutting section described earlier. The difference between the two is merely that the classic origin of the cutting section is primarily to organize the location of final fellings and the interval between contiguous coupes and not to organize subsidiary work.

5. *The periodic block*

The B.C.F.T. defines the periodic block as 'the part or parts of a forest set aside to be regenerated or otherwise treated during a specified period'.

Periodic blocks are most commonly used in uniform forests, that is forests composed of more or less uniform even-aged stands, which are regenerated naturally by one or more regeneration fellings, but they can also be used with artificial regeneration.

In application of the method a regeneration period is selected, namely the reasonable length of time in which a considerable area of mature forest can be regenerated. That time will depend on the species and local conditions such as the frequency of seed years, the need of seedlings to have overhead protection and the time seedlings take to get established. The more difficult the conditions for regeneration the longer will the regeneration period be. Then, if the chosen regeneration period is twenty-five years and the rotation is 150 years, the number of periodic blocks is six, each block having an area of one-sixth of the felling series. So by formula the area of each $P . B = A = FS \times RP/R$, where FS = area of the felling series, R = the rotation and RP = the regeneration period.

It is not necessary to make each periodic block compact and self-contained though that may be convenient. Nor is it at all essential for each periodic block to be permanently fixed and demarcated on the ground although at one time that was common practice particularly in France where the methods of using periodic blocks have largely developed. It is not even essential for all the blocks to be selected and the areas of those selected may be temporary and liable to be altered at each revision of the working plan. Consequently periodic blocks can be used and classified in three different ways which are described below.

(i) *The permanent periodic block method.* In this method the blocks are selected for permanent retention and are preferably compact, self-contained and equi-productive units. Each block contains the age-classes suited to its status so that together all age-classes are represented in equal amounts. Thus if the rotation is hundred years and the regeneration period is twenty years, P.B. No. I contains age-classes 81–100, P.B. II contains age-classes 61–80 and so on to P.B. V with age-classes 1–20. Each block is regenerated in twenty years in turn from I to V.

With this method the organization is simple and very practical. Different kinds of work are concentrated in different blocks, regeneration fellings in the oldest block (all in one compact area), tending and cleaning of regeneration in the youngest block and thinnings, graded by average age and therefore size of tree, in the other blocks.

But to this simple, rigid method there are serious objections. In the first place the method requires uniform and easy, favourable conditions. Damage from insects, fire or climate or failure to get regeneration in the planned period may severely disrupt the age-class structure. Variation in sites or species will result in varying growth rates so that stands will reach maturity at different times but will have to be removed and regenerated at the end of the planned rotation. Therefore for the method to be successful accidents must be rare, regeneration in a given time must be certain, site conditions must be uniform and there must be only one main species or several species which reach desired maturity in the same time.

In the second place introduction of the method will unavoidably demand loss and sacrifice in the first rotation. To form compact, self-contained periodic blocks, the original allotment will inevitably entail the assignment of unsuitably aged stands to the several blocks. Some mature stands will have to be allotted to young periodic blocks and some immature stands to the regeneration block, and so on. In consequence many stands may have to be felled and regenerated either before or after they fully satisfy the objects of management, thereby causing loss and current sacrifice in return for the attainment of a simple and practical normal forest in one rotation.

Consequently the method has been generally abandoned except for one or two forests which are favourably sited and constituted, such as the beech forests of the Forêt de Lyon in France where the method has been used continuously since 1856 with excellent results.

(ii) *The revocable periodic block method.* To introduce more flexibility the permanence and compactness of the blocks in the permanent periodic block method can be abandoned. Then stands can be allotted more suitably to their appropriate blocks and their allotment can be changed according to circumstance at revision of the working plan or at the end of a period. Thus a comparatively immature stand ravaged by wind or insects may be promoted to the regeneration block before its due time and more slowly growing stands may have their promotion to the regeneration block delayed.

Since allotment to blocks is liable to change it becomes unnecessary to allot the blocks intermediate between the regeneration block and the block last regenerated if the treatments of those blocks are similar. For example, if there are six periodic blocks and P.B. I is the block under regeneration and P.B. VI is the block just regenerated, P.B.s III–V could, if they are subjected to similar thinning treatments, be thrown together as an undifferentiated P.B. 'Intermediate', P.B. II, which might require special treatment preparatory to its regeneration, could still be selected and distinguished as a separate block.

There has been a tendency to combine intermediate periodic blocks in that way in India where many forests of *Shorea robusta* have been under conversion from selective fellings to the Uniform System for some forty or fifty years, such as those of the Saranda forest division in Bihar (Phillips, P. J., 1924, Mooney, H. F., 1937, and Sinha, J. N., 1962).

(iii) *The single periodic block method.* Clearly it is the selection of the regeneration block which is critical. In the working of even-aged forest not only does the more or less sudden change from mature trees to seedlings constitute an ecological and silvicultural crisis but the decision when to fell and regenerate greatly affects the degree of success in attaining the objects of management. Accordingly the single periodic block method evolved from the revocable block method, especially in France where it is widely used.

The term single, therefore, refers to the regeneration block which is selected, according to the merits of each stand (or compartment if that is the unit of working), for regeneration in a given time. The minimum size of stand considered will depend on the intensity of working and may or may not be in terms of fractions of compartments. The factors that influence the selection of stands for regeneration will be the combined attributes of silvicultural condition, economic factors and locality of the stand (felling and regeneration of a stand out of turn may be advisable for needs of utilization, protection and so on). The single or regeneration block will, therefore, travel or float by a somewhat indefinite course round the forest. Consequently the method is sometimes called the floating periodic block method.

Having selected those stands or whole compartments whose early regeneration is desirable, their final allotment to the regeneration block is decided in one of two ways.

In the first way the regeneration period is fixed arbitrarily in the same fashion as in the permanent periodic block method as a fraction of the rotation, for example $\frac{1}{5}$. The area of the single periodic block will then be $\frac{1}{5}$ of the whole felling series. The stands originally chosen for regeneration must then be re-examined and their number modified so that their area is $\frac{1}{5}$ of the felling series. The modification will involve either the rejection of some stands whose regeneration is less urgent or the inclusion of more stands whose regeneration had not at first seemed necessary.

The second way is the converse of the first. All the stands selected for regeneration are so allotted to the regeneration block. Their area is then calculated as a fraction of the area of the whole felling series, for example $\frac{2}{11}$. The time allotted for regeneration of the block will then be $\frac{2}{11}$ of the rotation.

Allowance for partial regeneration. In either way a complication arises in those stands whose natural regeneration is only partially complete. What actual area should be credited to the regeneration block? The difficulty is usually settled by proportion of basal areas. Then, if the basal area of a stand before natural regeneration fellings began was 140 hoppus square feet per acre and now that the fellings are partially complete is 35 hoppus square feet per acre or one quarter of the original, then it is assumed that regeneration is 75 per cent complete; 25 per cent of the area of the stand is therefore credited to the regeneration block, whose area is kept at the previously calculated total by reducing newly selected areas by an area equal to the unregenerated balance. *Consequently the delay in regeneration is continued so that in effect the rotation is lengthened in proportion.* In most circumstances, therefore, it is preferable to treat the unregenerated portion as arrears to be made good in the next period by merely adding it to the normal area of the regeneration block.

Discussion of Periodic Block methods

It is important to note that in the permanent, fixed periodic block method no variation of the planned rotation can be made for individual stands. In the revocable and single periodic block methods, however, the planned rotation does not apply to individual stands (or compartments if the working unit is the compartment) since each stand is allotted to the regeneration block at the discretion of the manager at each revision of the working plan. Therefore a slowly growing stand may be re-selected for regeneration after R+y years and a rapidly growing stand after R—x years; the planned rotation becomes the mean of many removal ages and is used only for general organization and accountancy. The consequent flexibility of the revocable and particularly the single block methods enable them to be used in variable site conditions, with several species, in localities where climatic or other damage is liable to occur and, being flexible, the methods are more easily adapted to changes in market demands.

In France, where the periodic block methods have been greatly developed in company with natural regeneration and the Shelterwood Compartment or Uniform silvicultural system which uses the compartment as a unit of treatment, the stand tends to be identified with the compartment even with the single periodic block method. Moreover in France, there has also been, until recently, a tendency to use only the indigenous species and either to give of them, such as beech in Lyon-la-Forêt and oak in Tronçais, supreme importance, and the others subsidiary cultural or weed values only, or to use two main species, such as silver fir and Norway spruce, which grow together amicably and have similar rates of growth. Large stand units can then give good results under comparatively extensive treatment by whole compartments. But the flexible single periodic block method can easily be used more intensively with small stand units, naturally or artificially regenerated, and several different species and several rates of growth, provided that the species grow in pure stands or have similar growth rates when grown in mixture.

In application of the single periodic block system (*affectation unique* in France), the regeneration block is painted blue on the map and may be called the *affectation bleu*. Of the remaining stands (or compartments) the older ones are painted yellow (*affectation jaune*) and the younger ones white (*affectation blanche*). That is simple and straightforward but there is another silvicultural method of management, the regeneration area method, in which the same colours are similarly used. Blue is again used for areas selected for regeneration; but in this method no time limit is fixed for completing the regeneration which is effected as expeditiously as conditions will allow. In France this area is termed the *quartier de regeneration* or *quartier bleu* in contradistinction from the *affectation bleu* of the single periodic block method in which a time for regeneration is fixed.

Consequently outside of France confusion has arisen by wrongly applying the term *quartier bleu* (with no time limit) to the single periodic block method (with its time limit for regeneration); the *quartier bleu* can apply only to the regeneration area method.

The basic difference between the single periodic block and regeneration area methods is the time limit for regeneration in the one and none in the other. Consequently different methods have to be used in calculating the permissible yields (see chapter VIII). Additionally the absence of a regeneration time factor and the more flexible type of yield prescription enable the regeneration area method to be used for more or less uneven-aged forests whereas the single periodic block method is suitable only for even-aged forests. In fact the absence of a time limit is itself likely to cause irregularity since the basic definition of an even-aged uniform stand is its formation or regeneration within a definite and comparatively short period.

6. *The felling (or cutting) cycle*

This term may be defined as the period in which all portions of a felling series are worked under a definite type of felling in a planned sequence.

The term must not be confused with 'rotation'; it has nothing to do with the production period, i.e. the interval between formation or regeneration and final harvesting of a crop. The most common use of a felling cycle is in uneven-aged forests worked under the selection system or selective fellings. The felling series is then divided into the appropriate members of sections, say ten, and fellings are confined to one section each year in turn. Clearly, the longer the felling cycle the smaller is the section to which fellings are confined in any one year; the fellings are, therefore, concentrated by long cycles and dispersed by short ones. In a selection forest if the annual yield to be cut is 2 per cent (and it may well be more) of the whole growing stock volume, then 20 per cent of the standing volume is cut in any one section on a ten-year cycle; on a twenty-year cycle it would be forty per cent—nearly half of the stands. The silvicultural implications of the length of felling cycle are therefore important but liable to be over-shadowed by the obvious utilization and sale aspects.

Long felling cycles, such as twenty-five to thirty years or more, which (often confined to only a few valuable species out of many) are frequently used in selective fellings in extensively managed tropical forests, may amount to one quarter (or even one-third) of the exploitable ages of many of the species. In effect there will then be, say, four age-classes in the forest— a new age-class appearing from regeneration at each cycle of felling up to exploitable age, the oldest being removed wholly or in large part.

The term polycyclic has been applied (Dawkins, 1958) to uneven-aged forests worked under a felling cycle as opposed to monocyclic, even-aged forests worked under a rotation.

It is also worth noting that in France the felling cycle is termed *rotation* and the rotation is called *revolution*.

The planned interval between successive thinnings of even-aged stands is called the thinning cycle.

C. ADMINISTRATIVE ORGANIZATION

1. The Sub-manager's charge

The organization of territorial administrative unit depends on the extent of the forests under one ownership and consequent size of the administrative and executive staff and their organization. But the basic unit on which the whole organization (such as that of a state forest service) is built may be taken to be the sub-manager's charge. The sub-manager, often called a forester or ranger, works under the lowest professional forest officer, who may be termed the district or divisional forest officer or local manager.

The forester or ranger will have a territorial charge, termed sometimes a range or forester's beat or, in Britain a section, which is also an accounting unit. The size of a range, beat or section will naturally vary with its compactness and the intensity of management. Thus in India the range may vary from 30–100 square miles in extent whereas in Britain the section is some 2000–4000 acres (3–6 square miles) of forest, so that some three to six ranges, beats or sections are included in a district or divisional officer's charge.

The range, forester's beat or section may itself be subdivided. Thus in India the range officers range may include two or more forester's beats which themselves may include two or more forest guard's beats for purpose of patrolling the forest and supervising silvicultural and other work. In Britain the forester's section may include two or more assistant forester's beats each of which contains three of four working blocks or groups of compartments (see above).

Thus administratively a forest is composed of one or several sub-manager's charges which are accounting units and form executive entities. These executive entities may contain sub-units in charge of assistants to the sub-manager for control and supervision of work. It is evident, therefore, that units of silvicultural organization should correspond with the sub-managerial range, beat or section. Thus if a forest contains four sub-managerial units, each working circle (unless very small and unimportant) could conveniently be subdivided into four felling series, one for each range, beat or section.

As the silvicultural organization of each felling series should be complete in itself—in silvicultural treatment, regulation of its yield and control of its outlines—its sub-division into periodic blocks or regeneration areas, felling cycle sections, cutting sections or working blocks can be regulated to suit both the growing stock and site peculiarities and the administrative peculiarities of the sub-managerial unit. As the latter is also an accounting unit, accounts for each felling series could also be easily separated.

2. Higher organization

The higher administrative organization of personnel and their responsibilities must obviously vary greatly with the type of ownership and the extent of forests that are administered. Thus there must be great differences between the administration of small private forests, large company forests and vast national forests for which the state itself is responsible.

There are three main principles to guide the organization of the administration of forests, principles which become the more relevant and important as the number of the administrators increase with the size and value of the forest estate. Their full application will attach to the administration of national forests which are not only large but have multiple uses and have wide social effects so that they require an elaborate organization.

The three principles are:

(a) There should be a clear chain of command with definition of responsibilities for each link in the chain from the bottom to the top where the chain should end in the owner or owning body which for state forests would be the government minister responsible to the elected house of representatives.

(b) The chain of command should include specialists or experts in the various branches of forestry including research and economics so that they can exercise their skills in more than a mere advisory capacity but without upsetting co-ordinated action.

(c) There should be adequate connections with other activities particularly those dealing with uses of land, consumers of forest produce and sources of finance.

The first principle is comparatively easy to observe and is exemplified by a forest service manned by trained forest officers in a government department responsible to a minister (see Table 8). The chain of command might be headed by a chief

Table 8

ORGANIZATION OF A FOREST SERVICE AND ITS RESPONSIBILITIES

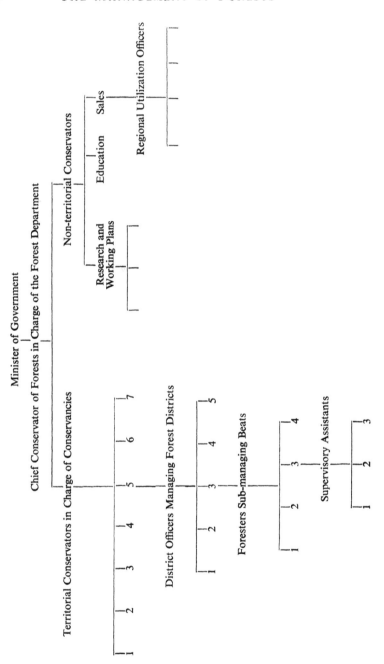

conservator (or director) of forests who is the head of the forest department and directly responsible to the appropriate minister. Under the chief conservator or director of forests would be a number of conservators each of whom would be in charge of the forests in a region of the country. Under each conservator there would be a number of deputy or assistant conservators each in charge of a forest or group of forests as its district or divisional forest officer. As we have seen earlier the district officer is the local forest manager under whom are the sub-managers or foresters or rangers in charge of a section, range or beat. In addition there might be one or more 'non-territorial' conservators in charge of general activities such as research, forest education, working plans and markets.

The disadvantage of such an organization is that administrators cannot themselves be expert in each of the many branches of forestry with which they must be concerned even if their previous experience has involved periods devoted to several of them. Technical advances and the greater social and industrial ramifications of modern forestry make it impossible for one individual to keep up-to-date in all aspects of his profession and deal with the mass of relevant detail of day to day administration. A chief conservator or conservator must, in consequence, have under him specialists to advise him so that his decisions are tempered by their advice. He may also delegate authority to them to execute decisions. Additionally he may come to decisions only after having held meetings with his specialists so that their advice, which may conflict with each other, can be co-ordinated. Nevertheless the final responsibility rests with the senior, titular head and the advice of his juniors, given without personal responsibility for final decision, may be coloured too much by a narrow specialist outlook.

To spread the load of work and responsibility, to introduce special knowledge and experience and to co-ordinate decisions and their execution at each stage of the administration a system of managerial boards of directors can be adopted. Each board would be headed by a chairman and have a number of managing directors each responsible for a particular main branch or aspect of forestry. An organization of that kind should bring into a government department the vitality of a big business concern but might encourage the disadvantages of having too much administrative personnel and of weakening the direct chain of command. But those two disadvantages really depend on the way the organization is used and on the calibre of the managing directors at each stage.

An organization of that kind has been recently introduced in the Forest Department in Britain. It is represented in Table 9.

It can be seen from the table that the board of the Commissioners includes a full-time administrator, who normally but not necessarily might come from the ranks of the civil service such as from the Finance Department, and two full-time experts on forest management and harvesting and sales who might normally but not necessarily be members of the forest service. Those three specialist posts are repeated at conservancy level. Co-ordination of their functions is assured at board level by the Chairman of the Commissioners and at territorial conservancy level by each conservator. Training and education and co-ordination with private forestry is included under administration and finance; working plans and research under management. So that at head-quarters (and in conservancies) there are represented the several branches of State forestry which correspond roughly with the non-territorial conservancies of the previous organization in Table 1 but are not necessarily held by members of the State forest service. It can also be seen that by the presence of representatives of private forestry, the timber trade and commerce on the board of commissioners those aspects of Forestry Commission policy are widened and strengthened.

As each of the main branches represented by a full-time director of the board of commissioners must be supported by an adequate organization of personnel it may appear that there is some danger of excessive development of establishment at headquarters. But the ramifications of State forestry, from recreation and private forestry to commercial forest management, are so wide that a heavy upper establishment seems inevitable. It would seem to be wise to ensure an accountancy procedure that separates charges so that overheads at headquarters and in the conservancies are not charged unnecessarily to the costs of managing the forests themselves and that legitimate charges on the forests are debited as appropriate to the separate main objects of management—recreation and amenity, social benefit and profit—and the units of forest allotted to them.

Table 9

ORGANIZATION OF THE BRITISH FORESTRY COMMISSION

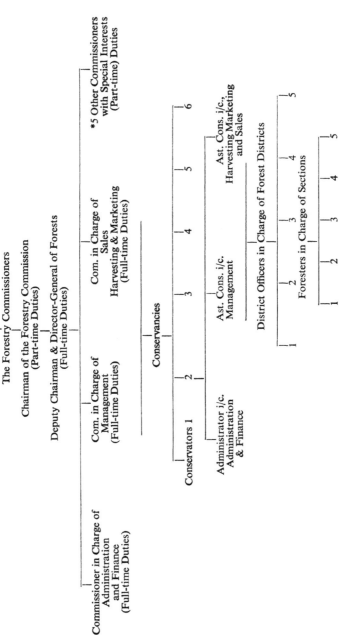

* *Note:* The special interests of these Commissioners are: Private Forestry—2; Trade Unions—1; Commerce—1; Timber Trade—1.

The obvious purpose of dividing a forest which may be large and scattered into component managerial parts is to promote efficiency in achieving the objects of management which may both be multiple and divergent in themselves and require divergent methods of working the forest. To prevent confusion and discontinuity of action subdivision aims at both concentrating separate spheres of work in separate localities and dispersing the work as uniformly as possible among the administrative and executive staff.

The subdivision of a forest, therefore, has the following main aims:

(i) Separation of areas that are treated differently for three main reasons:

(a) The objects of management are basically distinct so that in each separate part there is one over-riding object to which others are subservient.

(b) Different localities and growing stock, in combination with the objects of management, require basically different silvicultural methods.

(c) The development stages of the stands differ. The basic stages are formation and establishment, adolescent evolution and fruition of maturity.

(ii) Separation of areas for administrative convenience especially to:

(a) Equalize work among sub-managerial staff and labour.

(b) Facilitate losses and sales.

(c) Allow recovery periods for repairs to roads, drains and fences.

(iii) Separation of areas for convenience of markets in forests where topography, communication and means of transport confine supplies for different market centres to different tracts of a forest.

(iv) Separation of areas to facilitate the keeping of accounts and other records so that they can be used to diagnose results. The information wanted can be broadly classified as:

(a) Silvicultural methods—treatment of growing stock related to species, development stages and sites.

(b) Exploitation methods with which sales methods are allied.

(c) Maintenance methods—repairs and general preservation of the estate.

(d) General control and efficiency and costs.

(e) Functional activities.

Bibliography

Anon. (1953) 'British Commonwealth Forest Terminology', Part I. Empire Forestry Association, London

Anon. (1960) 'Working plan for Dean and Highmeadow.' Typescript for internal circulation, Forestry Commission of Great Britain

Anon. (1965) *Manuel Pratique d'aménagement*. Ministère de l'Agriculture, Direction générale des Eaux et Forêts

Dawkins, H. C. (1958) *The Management of Natural Tropical High Forest with Special Reference to Uganda*. Commonwealth Forestry Institute, University of Oxford, Paper No. 34

Eggeling, W. J. (1947) *Working Plan for the Budonga and Siba forests of Uganda*, first revision, period 1945 to 1954. Entebbe, The Government Printer, Uganda

Meyer, H. A., Recknagel, A. B., Stevenson, D. D., Bartoo, R. A. (1961) *Forest Management*, 2nd edition. The Ronald Press Coy., New York

Mooney, H. F. (1937) *Third Revised Working Plan for the Revised and Protected Forests of the Saranda Division, Bihar, 1936/37 to 1955/56*. Superintendent, Government Printing, Patna, Bihar

Osmaston, F. C. (1957) 'Yield Regulation,' *Quarterly Journal of Forestry*, II(2)

Phillips, P. J. (1929) *Revised Working Plan for the Revised Forests of the Saranda and Kolhan Division in the Singhbhum District*. Superintendent, Government Printing, Patna, Bihar

Sinha, J. N. (1962) *Fourth Revised Working Plan for the Revised and Protected Forests of Saranda Division, 1956/7 to 1975/6*. Superintendent, Secretariat Press, Patna, Bihar

Chapter VII

THE ACTUAL GROWING STOCK AND ITS INCREMENT

GENERAL MATTERS

The forester always has to deal with the growing stock and land in his charge as they actually are. It is their present condition, tempered by their potential productive capacity in relation to current and expected future markets, that influences all planning and action. Without knowing what there is in the forest, even initial exploitation for current markets cannot be efficiently organized; without knowing both the potential productive capacity of sites and the extent and type of future demand for forest produce, neither the long-term purposes of management can be decided nor can the stages of development and the rate of investment be arranged.

Although both the early difficulties of deciding the long-term purpose of management and the means of achieving the purpose may be highly complex and dependent on information hard to get in a forest which is in an early stage of development, the need for a full knowledge of the growing stock and its parent sites never diminishes. Indeed the need for detailed information grows as management develops and becomes more refined and intense; the need to relate changes and performance of the growing stock and sites with treatment and demand becomes more acute.

The management of forests, therefore, constantly depends on assessments of what the forest contains and how its constituents are behaving in order to decide what to do next and how to do it. The fact that the trees are themselves both the product to be harvested and, allied with the site factors, the machinery of production, complicates assessment. Those combined properties of product and produce, together with the dispersal and proportions of different types of growing stock, the variety of site factors and their effect on increment, influence and permeate all appraisals of a forest and its resources. Information collected must tell the forest manager on the one hand what products are available for the market, where they are and in what quantity and, on the other hand, how the productive machinery can be tended and renewed to produce cheaply more and better goods. The information is used for two main but interdependent purposes namely the silvicultural tending and management of the growing stock (i.e. of production) and the organization of harvesting and sale or other disposal of the products of that growing stock.

In his silvicultural management the forester must know how each part of the growing stock is performing relative to the objects of management and the potential capacity of the sites. Then he can organize fellings and treatment so that the existing growing stock becomes more productive of what is wanted and he can change the species to those more capable of using the site potential and satisfying markets. If the objects of management include sustained yields the forester must also know what the relation between the present and normal state of the age or size-classes are so that he can adjust them.

The silvicultural manager must therefore know not only the constitution of the growing stock—in specific composition, structure and age or size-classes—but also the density and increment of each part or stand of the forest. In addition the silvicultural manager must assess the site factors everywhere— the soil and its hydrology, the microclimate—so that he can establish the capacity of each site and the means to improve it.

Silvicultural management is always concerned with a long-term view which influences all present action. But the organization of harvesting and sales, i.e. the utilization of the forest, is concerned mainly with a short-term view of what produce is available now and in the near future although the long-term view must also play an important part in planning. That is particularly true in regions of afforestation and management of undeveloped forests where new wood industries may have to be introduced when the forests become fully productive.

The utilization manager therefore is interested mainly in the quantities, qualities and sizes and position of forest produce which are expressed by the specific composition, density and size or age-class constitution of the several stands of the growing stock. But that inventory of what exists does not necessarily mean that the utilization can take what it is convenient to take where and when it is convenient. Removals must be tempered by and be agreed with the silvicultural management. This essential and obvious link between the two sides of management is still more important in forecasting and planning for future yields which will always depend on the success of silvicultural management and the consequent product, its increment and sequence of maturities. The silviculture and utilization must be partners dependent on each other on an understanding that dominance of utilization, particularly where demand exceeds supply, is liable to endanger the safety of the growing stock and continuance of supplies.

Therefore very complete inventories and assessments of the growing stock and sites are necessary and must be frequently repeated to show what changes occur as the result of management. Although the types of information wanted are interdependent and entwined with each other they can be considered under three main heads—constituents of the growing stock (structure and specific composition, density, ages and sizes and hence volumes and values); increment; properties of the sites.

1. SPECIFIC COMPOSITION, STRUCTURE, VOLUME AND VALUE OF THE GROWING STOCK

The means of collecting the obviously essential information of the contents of a forest depends on the science and techniques of forest mensuration and will not be elaborated here. The reader should consult works on forest mensuration such as those of Spurr (1952), Bruce and Schumacher (1942), Pardé (1961) and Husch (1963) and works on photogrammetry and interpretation of aerial photographs such as Spurr (1960). Here we are concerned with the type of information wanted for management.

Basically the forest manager wants to know what species grow in the forest, where they are and what their volumes and degree of maturity are in the several places.[1] In addition he must know what are the composition and structure of the stands, namely whether the species are pure or mixed together and whether the stands are coppice or high forest, uniform and even-aged or irregular and uneven-aged. Those factors of specific composition

---

[5] He also wants to know their rates of growth; this will be discussed under increment below.

and structure of stands enable the growing stock to be classified, delineated and mapped in a series of forest types which form a base on which to build the main design of management. The forest types, which also tend to express the factors of sites, influence decisions on both the purpose of management and methods of treatment. Thus bamboo and teak forest types, or pine and oak types, will be treated differently and supply different markets.

Forest types

A management forest type may be defined as a class of forest vegetation that is sufficiently distinct to require a different kind of management or treatment from that of other forest types.

In virgin or undeveloped forests the management forest types will coincide at least with broad ecological site-types although small changes in site-type and consequent small changes in the vegetation would be ignored unless the change requires a different management policy or technique of treatment. In fact in Finland and elsewhere the ground flora has been used, after the proposals of Cajander (1926), for the classification of forest types and their variations in growth potential. In developed and artificial forests also there should clearly be a close connection between forest and site types since the efficient and ideal forest will contain species sited to the best advantage and grown in the way most suited to the conditions (within the limits of the objects of management).

This utilitarian classification of the growing stock by forest types has also been used in the U.S.A. (Chapman, 1950) as a basis for the formation of working plan areas and working circles (working circles and working groups respectively in the nomenclature of the U.S.A.). But in spite of this utilitarian aspect of forest types it should be remembered that their value and purpose is to classify and map the growing stock as *it is*, and the existing forest types may not tally with the working circles actually made. It may well be that part of a forest type distinguished by being, say, even-aged Douglas fir may be found on sites unsuited to Douglas fir so that it will be placed in another working circle managed, perhaps, for pine pulp-wood to which the Douglas fir will be converted in due course.

Forest types, therefore, are differentiated firstly by structure— whether the stands are coppice or high-forest, and whether the high forest is uniform and even-aged or irregular and uneven-aged. Secondly, within one of those three primary structural classes, subdivision would be by species and thirdly, within a specific type, by quality or site values as necessary. In essence, therefore, the differentiation of the growing stock by forest types resolves into a stock-mapping of the forest by species according to forest structure and, if necessary, also by quality or yield class which varies with site peculiarities. A simple example might be high forest containing both uneven-aged and even-aged growing stock; the uneven-aged stock might be composed of beech with some subsidiary oak and the even-aged stock might contain three specific types, Scots pine, Douglas fir and Norway spruce. There would then be four main forest types, one of which was uneven-aged. If any of those types were to include a quality so different from the rest of the type that it would require separate treatment it should be divided into two sub-types. Thus in the above example there might be stands of Douglas fir so poor in quality that their conversion to, say, Scots pine is desirable. Two Douglas fir types—a good and a bad one—would then be distinguished.

Difficulties arise in distinguishing forest types among mixtures so that arbitrary definitions of purity are necessary. In Britain the practice is to regard a crop as pure when a species occupies 80 per cent or more of the canopy; if there are only two species in a mixture only those occupying 40 per cent or more of the canopy are mentioned; if there are three or more constituents those occupying 30 per cent or more of the crop are mentioned (Anon., Working Plans Code, 1965).

Stand types

Forest types are too varied in themselves to give adequate information or provide acceptable populations for sample enumerations except perhaps in hitherto unmanaged or very extensively managed forest. Thus an even-aged pure Scots pine forest type may contain stands of very different ages from seedlings to overmature timber trees. It is obviously necessary to divide each forest type into stand types which require different treatments or have outputs of different types or quantities of produce.

A stand type may be defined as a collection of stands within a forest type which are sufficiently similar to require the same treatment and planning attention or produce the same type of produce.

There are four main features that influence the selection and separation of stands and their allotment to stand types within a forest type, namely:

(a) *Extent of stand.* In intense management of valuable forest, stands as small as one or even half an acre may have to be recognized whereas in extensive working of low value forests, stands of twenty or more acres may be ignored (see also sub-compartments in chapter VI).

(b) *Age of stand.* In even-aged forests one must know and therefore must map the area of each age-class. How narrow should the age-class interval be? Mere difference in treatment is inadequate to decide the interval. The effect of different age-classes on general planning and the type or quantity of out-turn from them is equally or more important. Thus in Britain Sitka spruce between the ages of twenty and forty years (on a rotation of some fifty-five years) may be subjected to the same thinning treatment. But an age-class interval of twenty years is inadequate for forecasting accurately either present or future yields or for planning final fellings to regulate the sequence of maturing timber. A much narrower interval is needed.

For intensive management of conifer forests in Britain therefore, the recognizable age-class interval must be as narrow as five or at most ten years. For slower growing species worked on long rotations and for more extensive treatment, for example where thinnings are unsaleable, intervals of fifteen or even twenty years are small enough. Thus for oak managed on a rotation of 120–150 years in Britain age-classes of fifteen years suffice.

(c) *Treatment of stand.* An alternative method of classifying stands is to use development or treatment classes instead of age-classes. Thus there might be six development classes composed of the seedling and thicket, young pole, large pole, young timber, mature timber and overmature timber stages and six equivalent treatment classes.

An advantage of such a classification is that it expresses more truly the degree of maturity of a stand than does its age. The time at which a stand becomes mature depends on its rate of growth which varies not only with species but also with quality of site. Thus a sixty-year-old stand of Sitka spruce may be fully mature but a sixty-year-old stand of oak may be only half or one-third mature. But a development or treatment class correctly indicates the degree of maturity, not in absolute time but as a percentage or proportion of the time taken to reach maturity. The actual time taken by a species of a certain quality class to pass through a development class will differ from that taken by another species or quality, but the ratio of the time spent in a class to its final felling age will be the same. Thus sixty-year-old oak might still be in the young pole stage when sixty-year-old Sitka spruce is mature timber.

An example of classification by development and treatment classes and their approximate degrees of maturity is shown below:

Development class	Treatment class	Approximate % of maturity
I Seedling and thicket	Weeding and cleaning	0–20
II Young pole	Early thinning	21–40
III Large pole	Middle thinning	41–60
IV Young timber	Late thinning	61–80
V Mature timber	Regeneration felling	81–100
VI Overmature timber	Salvage felling	Over 100

It is clear from the table that if 20 per cent of the area of the forest is found in each of the first five classes the forest is 'normal'.

Another advantage of the method is in forests where the age of stands is not recorded and cannot be known, e.g. where the species has no annual rings. A subjective assessment of the treatment classes of stands would, however, be possible. Moreover freedom from an age outlook invites flexibility so that an unhealthy or excessively open stand can be allotted without hesitation to regeneration felling or even salvage felling whether it is twenty, fifty or a hundred years old, although to do so will require special action in forecasting the timber outturn. But reasonably accurate allotment to the middle classes is not easy; but the allotment is not critical until entry into class IV (see p. 185 *et sqq.*).

(*d*) *Health and density of stand.* Areas which have been damaged by wind, fire, insects or other adversity so that their density or future prospects demand special attention (such as clear-felling and regeneration or enrichment by planting) should be classed as separate stands. This separation is necessary both because of the different treatment needed and because of differences in standing volumes and yields per acre.

Having stock-mapped the forest, by ground survey alone or with the help of aerial photographs, the position and extent of each forest and stand type is known together with a description of them, compartment by compartment and for the forest as a whole. For an early stage of extensive management that information is adequate at least to start management and make preliminary plans. But a knowledge of volumes and increment soon becomes essential. The growing stock must be measured.

2. VOLUMES AND INCREMENT

As volume and increment are so closely related and are usually measured at the same time the management implications will be discussed under increment below.

The increment

The increment of a forest is its most vital character. It is intimately connected not only with the health and species of the trees and the fertility of the sites that they occupy but with the volume and age of the stands. A virgin forest, or any forest that has been undisturbed for a long time, has no net increment; in fact its increment may even be slightly negative for a period succeeded by an equivalent positive increment. Individual trees may be growing fast or slow but others decay and die. The volume of the forest vegetation has reached a climax more than which the site cannot sustain. The forest has achieved a static equilibrium. An example given by Chapman (1950) is reproduced in Table 10, with kind permission of the Yale Cooperative Corporation, New Haven, New Haven, Connecticut.

But, if the virgin or long neglected forest is used by cutting its timber it will try to regain the climax volume; it will then have a positive net increment.

In a fully and efficiently managed forest the volume of the growing stock is comparatively small but its net annual increment is high (some 2 to 7 or 8 per cent of the total volume for high forest) the amount depending on the age to which the trees are grown as well as on the species used and the fertility of the site. The continued yields from the forest depend on that net increment so that it is essential to attain that dynamic state of equilibrium which provides the maximum increment of what is wanted. That state of equilibrium is achieved and maintained by cutting and by regeneration of the mature trees cut.

Table 10

Increment, Virgin Forest, in absence of cutting—fifty years—on 438,423 acres. Ponderosa Pine, Coconino National Forest, Arizona

| Class | Age | Increment per acre For 50 years | | | Increment on total area | | | |
		Loss Bd. ft.	Gain Bd. ft.	Area acres	For 50 years Loss	Gain 1000 feet	Per year Loss B.M.	Gain
Veteran	300	5,630		163,590	921,012		18,520	
Mature	200		530	36,469		19,329		386
Blackjack	90		5,360	91,072		488,146		9,763
Poles	50		7,480	37,486		230,395		5,608
Seedlings	20			109,806				

Total loss per year　　　　　　18,520 M. ft. B.M.
Total gain per year　　　　　　15,757 M. ft. B.M.
Net loss per year　　　　　　　2,763 M. ft. B.M.
Net loss per acre, annual, approx. 6 ft. B.M.

The present volume and value of a hitherto unmanaged forest will decide its initial exploitation, that is to say the use and disposal of accumulated capital. The continued management of the land, for forest products or something else, will then depend on the amount, quality and value of the timber or other product that can be annually expected from it; the manner of working the land as forest will depend on how the increment can be maintained or improved in quantity and quality in such a way that utilization of it is promoted.

Effects of increment on policy

The amount and quality of annual increment expected from forests have far reaching effects on decisions of policy such as those concerning:

(*a*) The area of forest required either to supply a country's needs or to satisfy local industrial or other demands.

(*b*) The types of land that can be economically afforested or maintained under forest.

(c) The kind of wood-using industry that can be supported.
(d) The size of industrial plant that can be installed.
(e) The kind of exports that can be developed and their extent.

Effect of increment on working methods

In commercially managed forests where the object is to provide material goods rather than aesthetic or protective and other service values, the rate at which the goods are produced is the basic element in the economy of the forest. The rate of growth or increment of the growing stock is the earning power of that growing stock or capital. The forest manager is always concerned with getting the greatest volume of valuable increment from the smallest volume of growing stock consistent with the objects of management.

Increment must therefore greatly influence the manner of working a forest in the three interdependent aspects given below:

(a) Choice of species.
(b) Choice of silvicultural system and methods of tending.
(c) Choice of the times at which to fell and regenerate.

(a) *Choice of species.* It is obvious that if two species were to grow equally healthily on a site and produce equally desirable and valuable timber the one that grows the faster (in volume per acre) would be preferred. The choice is seldom so simple.

To compare the economic value of two species it is necessary to calculate the values of the yields of each at the respective times when they will be harvested and then discount the yields to the beginning, i.e. time of formation, and compare results. The one providing the greater net discounted revenue (NDR) at a chosen rate of interest or the highest financial yield is economically preferable. For such a calculation it is necessary to have reliable yield tables which show the expected volumes of the growing stock and volume of thinning yields at successive times in the rotation of each species.

Given the data of suitable yield tables (and accurate forecasts of timber values and costs of growing and harvesting) the calculation is comparatively straightforward provided that the species continue to behave as the tables predict. But early promise may not be fulfilled in the later life of the one species or may not be repeated in its succeeding generations but may be sustained in the other. Continued study of the increments and prospects of every stand for more accurate assessments of the continued performance of species in various sites is obviously necessary.

(b) *Choice of silvicultural system and methods of tending.* Disregarding the species involved and their intrinsic characters of growth, the amount and rate of vegetation produced, preferably judged by its dry weight, depend on the fertility of the site. That fertility depends on adequate and suitable supplies of water and nutrients in the soil, carbon dioxide in the air and solar heat and light.

Bearing that in mind, the forester is concerned with three main aspects which can be influenced by the silvicultural system and method of treatment that he may adopt to obtain the best growth for his purpose. Those aspects are:

(i) Improvement and greatest use of the fertility of the site.
(ii) Confinement of vegetative growth to trees.
(iii) Confinement of tree growth to the most valuable kinds of wood.

(i) *Site fertility.* The fertility of a soil depends on its depth, its chemical and physical constitution and water content. Deficiencies of certain chemical elements such as nitrogen, phosphorus and potassium are quite common (Hewitt, 1966) and deficiencies of others, including trace elements, not rare. Such deficiencies may have a serious effect on growth but once their need has been established they can be physically applied to the soil in fertilizers. This application of fertilizers is common in Britain for afforestation of peaty soils and the possibilities of manuring older trees is being tried in several countries, e.g. New Zealand where aerial application of phosphate fertilizers on *Pinus radiata* has been tried (Conway, 1962). But as Binns (1966) has pointed out fertilizers will increase weed growth which choke young trees so that a combination of fertilizers and herbicides may be necessary for young regeneration.

The virgin forest returns to the soil all nutrients so that site fertility is improved by root penetration with consequent physical aeration and fragmentation of compact soils, an improvement encouraged by the natural addition of most micronutrients (Hewitt, 1966) such as Fe, Mn, Cu, Zn, B, Mo,

CO and Na which can occur in atmospheric precipitation. In contrast, the soil in managed forest suffers a continued loss of nutrients by the removals of timber and other forest products. The managed forest may also endure periodic crises when a mature stand is clear-felled resulting in a marked reduction in transpiration and exposure of the soil causing a rise in the soil water level and desiccation of the soil surface. The ecology of the site is drastically changed with probable adverse effects on tree growth.

An example which severely taxes the soil is coppice with standards. The coppice is frequently clear-felled and removed, and the removals usually include most of the brushwood and bark which are the main stores of nutrients (Rennie, 1957). The standards have their crowns isolated and so are exposed to sun and air currents and consequently to greater transpiration although they have to compete with the coppice for both water supplies and nutrients. The standards tend to suffer from drought. The system therefore needs deep, fertile soils in a drought-free climate and may need periodic doses of fertilizers to ensure good, continued yields.

As mentioned above soil fertility is greatly affected by the hydrology of the soil. The matter is complex and is well summarized by Rutter (1966) who points out that although growth in the earlier parts of the season (e.g. of conifers in Britain) is not likely to be affected by drought since it occurs before a marked water deficit can have developed in the soil, there may be delayed effects in many species. In many species the new shoots and leaves are preformed in the bud at the end of the summer and the effects of drought then may be seen in the following year. Rutter (1966) also refers to the way in which soil water affects nutrition and therefore growth. If the surface soil dries, growth is restricted not because the water balance of the tree is materially affected but because its uptake of nutrients is affected.

A markedly fluctuating water balance may endanger the water balance of trees, particularly in badly aerated soils in which lower roots may die at times of high water levels, e.g. in winter or early spring, and subsequently the trees suffer from an inadequate root expanse when the soil dries. The stagheadedness frequently seen in mature oak in France after a heavy seeding felling may be the result of a lack of aeration caused by a pronounced rise of the water level as the result of the reduced transpiration per unit area although the individual and isolated mother trees themselves are exposed to greater transpiration than before. In places where conditions are difficult for regeneration (especially natural) such as on southern aspects where spring droughts may occur, particular attention to the mode of making final fellings may be necessary. Slim strips, running from NE to SW in the northern hemisphere and cut successively in a southerly direction, may be needed to give protection from the sun, strips which are kept narrow in width so that there is no material reduction of transpiration. The water level of the soil is then not unduly disturbed and side shade reduces desiccation of the seedlings and surface soil.

Knuchel (1953) stresses the importance of carbon dioxide in the air by pointing out that the world's plant covering absorbs such large quantities of carbon that the supplies of carbon dioxide would be exhausted in three years if it were not replenished—mainly by decaying plant detritus. He implies that carbon dioxide may be in short supply locally when the air is still. A growing stock composed of several tiers of trees, such as in a selection forest, would then be better able to use the available carbon dioxide than an even-aged growing stock whose trees have their crowns at one level.

Moreover Knuchel also suggests that the more isolated and therefore larger, longer crowns of the bigger trees in selection forest are both free from mutual damage caused by swaying and slashing in the wind and more capable of using available light than the crowded crowns of even-aged forest. Therefore the volume increment of selection forest, which does not suffer from periodic ecological crises, should not only be greater than that of even-aged forest but the increment is particularly favoured on the large more valuable timber trees whose crowns are free; the seedlings and saplings develop under and along with the timber trees so that no unit of land is confined to crop establishment but is always also producing timber increment. Köstler (1956) also supports the growth advantages of irregular growing stocks with their vertical density and compares even-aged stands with their simple horizontal density to 'an empty hall through which the wind blows'.

Unfortunately it is difficult or impossible to get truly comparable figures to compare the growing stock volumes and increments of the two basic silvicultural systems. Knuchel (1935), however, gives a comparison in a table showing volumes of growing stock over 5 inches quarter girth at breast height calculated from yield tables for clear-felling high forest and volumes found in selection forest. His figures (for mixed silver fir—Norway spruce—beech) are reproduced in Table 11, with permission from Oliver and Boyd, Edinburgh.

Table 11

COMPARISON OF VOLUMES IN CLEAR-FELLING AND SELECTION FOREST

Type of Forest	Volumes/acre, hop. ft.		
	I	II	III
Clear-felling high forest	6513	3773	1617
Selection forest	6175	4490	2800

The figures suggest that in the more fertile sites the selection forest has no advantage but has an advantage where site fertility is low or poor. It may be inferred that selection forest is better able to use site factors, a capability which is not exercised in fully fertile sites.

Even if that conclusion is true it remains true that only a few species are suited to the selection system. In Europe only silver fir and Norway spruce with beech as a subsidiary are fully suited. Ideally the species should not only be highly tolerant of shade but should have crowns narrow relative to their stem width, i.e. a low crown/stem diameter ratio, without which not only does each tree occupy much space to the hindrance of those below but its felling will cause heavy damage. Dawkins (1958) has shown how serious is the damage caused by selective fellings in the tropics where the felling of a single climber-free tree of 8 feet girth destroys 0·05 acre of pole and adolescent stock; extraction by tractor increases the damage. Thus the felling of one large tree per acre would destroy perhaps forty years of growth over 20 per cent of the area.

Even when the forest manager rejects the selection system because he doubts whether the possibly greater productivity of the system is worth the dispersion of fellings, the greater felling damage and the generally more difficult working, he may still wish to design a growing stock structure which gives side-shelter from sun, drying winds and storm to young seedling, sapling and pole stands. Modification of the stand structure of the clear-felling system may be desirable particularly in rigorous conditions. Modification in the form of narrow strips or wedges may be desirable or an amalgamation of methods such as is found in the irregular shelterwood or *Femelschlag* system in which treatment is varied with site and species in one and the same forest.

What is certain is that treatment must suit the species and site to get the greatest production. It is now becoming easier to modify the environment or site, by applying fertilizers, by drainage, by irrigation and even, within limits, by changing the local topography with bull-dozer or dragline machines. Therefore it is becoming more important to understand the physiology of plants and trees. Unless it is known what factors are limiting or encouraging growth and to what extent, whether it be soil nutrients, water supplies, microclimate or even carbon dioxide, advances in methods of treatment in each forest can only be empirical.

(ii) *Confinement of vegetative growth to trees.* It is obviously desirable to limit the growth of vegetation other than trees if that other vegetation is competing for limited resources of the site. It is also desirable to supplement or replace a tree species with one or more others if the one is unable to use the resources of a site fully and thereby allows other vegetation such as herbs and shrubs to occupy the site. Examples are:

(*a*) In very young crops of trees the herbaceous flora, especially gregarious types such as many grasses, bracken or heather may compete for limited supplies of water and nutrients or, in a fertile soil, shroud young trees from light. Removal or killing of competing weeds is the obvious but probably expensive remedy.

(*b*) An older crop of trees, particularly those so-called light demanders and pioneer species, such as Scots pine, which form comparatively open stands, may not be able to use all the resources of a site—e.g. light. An understorey of shrubs and herbs will then develop. Provided there is plenty of water and soil nutrients the growth of the trees will still remain at the maximum of the species; a greater production of wood could then be obtained by introducing an understorey of a more light-tolerant tree species to take the place of the herbs and shrubs.

Alternatively an irregular system of silviculture might be used whereby regeneration of the tree species is encouraged at all times, in groups and patches and under the parent trees as in the Dauerwald and selection systems to replace weeds with young trees.

(c) Blanks, i.e. areas devoid of trees, should obviously be stocked.

(d) Stands may be thinned heavily to encourage more rapid growth of individual stems. Provided that the weed growth that results does not compete for limited supplies of water or nutrients no harm is done; the canopy soon closes again. But in regions where surface soil in particular tends to get dry, a herb layer of grass or bracken may affect tree growth. A greater production of wood per acre might then be obtained by lighter thinnings though a particular size of tree would be produced more slowly. If the limiting factor is water at depth, deep rooting weeds would need restriction.

(iii) *Confinement of growth to the most valuable kinds of wood.* Obviously in most circumstances the forest manager will wish to grow the more valuable rather than the less valuable kinds of wood. To a great extent the choice is determined by the site and the species that it can support. But within a site and species used, the method of treatment will affect the type, quality and form of wood and probably also the time in which the final product is provided.

Typical examples are:

(i) Some species, especially hardwoods such as beech or oak, have a heavy crown which in general is useful only for fuel and has a much lower value than the stem timber. To increase the proportion of stem to branch timber is clearly desirable. But to do so requires light thinnings and a crowded canopy which causes slow diameter growth of individual trees so that the rotation becomes economically too long. A careful balance of thinning intensity has to be struck.

(ii) The speed of growth, exemplified by width of annual rings, may affect timber quality. Thus in France the best quality of veneer oak is considered to be that whose annual rings have an even width which averages only 2 millimetres. Again, in conifers there has been much controversy on the effect of ring width on wood quality (see Aldridge and Hudson (1955),

Rendle and Phillips (1958) and Phillips (1962)) with arguments stating on the one hand that excessive ring width reduces density and strength of coniferous timber and on the other that it is juvenile wood, irrespective of its ring width, that is weak. This is not the place to explore the controversy or summarize conclusions. What must be stressed is that it is essential for the forest manager to know whether and to what extent speed of diameter growth affects the quality of the timber that he is growing. He can then adjust the thinning intensity so that the quality of timber is co-ordinated with the time taken to grow it so that the product is the most economic one (Hiley, 1955).

(iii) Knots usually reduce the value of timber but can be eliminated by timely pruning or be greatly reduced in many species by maintaining a dense canopy. To keep the canopy dense by subjecting the stand to light thinnings reduces the early returns and lengthens the rotation thereby increasing the cost of production. Pruning is comparatively expensive and must be done early and probably several times. The forest manager must compare the discounted costs involved with the discounted increased prices expected for the knot-free timber.

(iv) The form and straightness of trees influence greatly the quality and value of timber. Crooked, leaning trees contain compression and tension wood so that internal stresses affect its strength and liability to warp and split. The need to remove crooked, leaning and branchy trees in thinning and to space the remainder uniformly so that their boles grow as round and straight as possible is essential.

It is clear that thinnings have a great influence on the production of timber of quality. To decide on the intensity and possibly also the type of thinning, the forest manager depends greatly on the wood anatomist and his researches on wood structure and the basic causes of defects in timber as well as on a knowledge of the increased prices to be expected from high quality timber and the cost of producing it.

(*c*) *The time at which to fell and regenerate.* In an even-aged growing stock choice of the time when a stand should be finally felled and regenerated presents a continuous and vital problem to the forest manager. In an uneven-aged growing stock there is the same problem for each individual mature tree but it is entwined with the simultaneous cutting of immature trees as thinnings since it is not possible to differentiate between final and intermediate yields except by an arbitrary and unsatisfactory size limit. But one can say that the choice of a general exploitable size in an irregular growing stock is similar to the choice of a guiding rotation for an even-aged growing stock; in each type of growing stock it remains necessary to decide each year which actual stands or individual trees are to be finally felled, whether they have or have not reached the guiding exploitable size of rotation age.

Decision on when finally to fell a stand depends on the degree to which it is fulfilling or will fulfil the objects of management. If there is a single object of management, such as attainment of maximum NDR, the decision is comparatively easy although measurements and some calculations will be necessary. But if there are more objects than one they may conflict with one another thereby making a decision difficult unless an order of priority of the objects is prescribed. For example a sustained yield may be one object and maximum volume production may be another; to obtain a sustained yield requires the age-classes to be adjusted to conform with the normal distribution; to do that will usually demand the final felling of stands either before or after they reach the age of maximum volume production.

The actual state of a stand—its volume, health and increment—will always affect a decision on its final felling. Even if a sustained yield is the paramount object of management it would clearly be unreasonable to delay the felling and regeneration of a derelict stand of decaying trees when to fell it and restore the land to full production at once would merely delay normality for a slightly longer time. If the first object of management is maximum volume production it is clearly essential to know the rate of volume increment so as to be able to decide whether a stand has reached the peak of its performance, i.e. its maximum M.A.I. If maximum NDR is the main or only purpose it is necessary to know not only the present volume and timber prices of a stand (its present worth) but also its volume increment and future prices so as to be able to calculate the stands future value and discount that value to the present. Additionally the forest manager must know the prospective value of the stand which will replace the present one and discount it to the present. Then the combined present values of NDR of the existing land, felled now or five, ten or more years hence, can be compared. The time of felling which gives the greatest

NDR can then be selected. An example of the calculation, reproduced with permission from the Working Plans Code (1965) of the Forestry Commission of Britain is given in Appendix III.

It is evident that for his assessment of the growth potential of even-aged stands the forest manager relies greatly on yield tables. Although standing volumes and current rates of increment can be measured without great difficulty, projection of that rate of growth for more than a few years is unwise without the guidance of yield tables. Similarly or more so, forecasts of thinning yields, benefit from yield table figures. But conversely even when a stand is placed with confidence within a yield or quality class of a yield table at a known age the subsequent performance of the stand, particularly in youth, may not tally with the yield table. It may change from one to another yield class owing, perhaps, to its being in a site which cannot support the greater water or nutrient demands of an older crop. A constant check and maintenance of past records of the standing volumes, rates of increment and thinnings are necessary, preferably stand by stand.

Yield tables do not apply to an uneven-aged growing stock so that the forest manager then has to depend on his own measurements and records of performance. Moreover since age and size-classes, and probably species, are intimately mixed, stands cannot be easily separated; units of area such as the compartment take the place of the stand. Continuously repeated and comparable inventories, as practised in the method of control, compartment by compartment become the essential substitute for yield tables to show the results of past treatment and to suggest future performance. Moreover the compartment becomes the sustained yield unit since sustention acquires a great local importance not merely to provide regular local yields but because the whole structure and perpetuity of a true selection forest depends throughout the forest on the continued maintenance of stable proportions of the size- (and age-) classes.

3. SITE PROPERTIES

Without a full knowledge and assessment of the properties of the various sites in a forest and their relation to tree growth the management of the forest is severely handicapped. Basically the forest manager must know not only what species are growing in each site but what species *can* grow there and how well they will grow. He should also know what the deficiencies of the site are so that he can devise means to improve it.

The study of sites and their allotment to site types involves a complicated study of the several site factors which interact with each other to a lesser or greater degree. The main factors involved are the general climate, the topography and local or micro-climate, the geology, the subsoil, the soil and its water content. The vegetation, by its specific composition and vigour will express and indicate site conditions and the degree of fertility and thereby help in the classification of sites and, by comparison with other sites occupied by similar species, also help the understanding of causes for site differences. Vegetation which has been artificially introduced and cultivated is a site indicator only by its vigour and reproductive ability but at the same time may disturb and interfere with or even eliminate the occurrence and growth of natural vegetation whose value as a site indicator is thereby disguised or lost. The fauna will also indicate at least major site differences but owing to the mobility of many species they are not such definite and precise indicators of site conditions as vegetation. On the other hand their mobility enables them to occupy sites which become favourable much more quickly than vegetation. Moreover, if the number of one or more species becomes great, especially when introduced and bred by man, the fauna may limit the vegetation and even affect the properties of the site—for example, by destroying the vegetation and exposing the soil to erosion.

The influence of man on the vegetation and the site which it preserves should always be borne in mind. His use of herds of animals combined with methods of cultivation which loosen and expose soil to wind and water erosion and his use of fire destroying still more elements of protective vegetation must never be forgotten. Research into the history of man's use of the land as well as study of past climatic or geological changes may enlighten investigations into the causes of site infertility more than soil and water studies and give a clue to the future treatment needed.

The assessment of sites and their values, therefore, begins with mapping the vegetation of which the major feature will be the forest types which are also mapped for the management assessment of the growing stock. In virgin or long untouched forest, the forest types will have their natural associated shrub and ground flora all of which together express and reflect changes and values of sites. But the very great majority of forests have been used or exploited by man in comparatively recent years. An artificial tree population may have been imposed on the land or secondary growth may invade naturally after man's attacks on the forest; a secondary growth which will in a series of successions slowly revert to the previous climax forest if the forest is left untouched and if the exploitation had not been so severe and prolonged that the sites themselves were affected. In most forests therefore, particularly artificial forest, the shrub and ground flora, whose life cycle is usually much shorter than that of trees, will give better indications of site values than the trees.

A ground survey of the vegetation, aided by studies of aerial photographs, combined with an assessment of topography (slope, aspect, drainage pattern) and consequent microclimate should make it possible to decide on the site types needed and to map them even without a simultaneous soil survey—unless the vegetation has been so destroyed that it has no longer an indicative value.

A soil survey will then not only check and improve the preliminary site classification but should provide reasons why sites in a similar topography differ and indicate what species are likely to succeed. Reasons for lack of fertility will also show what can be done to improve sites—by drainage, application of fertilizers or even by irrigation and interference with subsoil and underlying pans of clay or other impervious material.

The scale and detail of ecological surveys and delineation of sites and assessment of soils will depend on the degree of development that has been reached in the management and on the value of the forest and its markets and consequent income which together decide the intensity of management. In the most highly developed and intensely managed forests of Europe the smallest recognizable sites are unlikely to be less than half an acre and are usually more than one acre. Exceptionally in very artificial, intensively cultivated and valuable poplar plantations, sites smaller than half an acre might be desirable. But in undeveloped, tropical forests with comparatively useless growing stocks any site under 25 or even more acres might be unworthy of recognition.

Similarly the classification of site types may vary from the very simple (good, fairly good and poor, for example) to the complicated one with five or six main types each sub-classified. The basis of classification may vary from one based on floral associations (quercetum, fagetum, pinetum, etc.) to one based on topography (valley, ridge, terrace and slope types) with emphasis on water supplies (water-receiving or losing sites) and sub-divisions based on soil peculiarities. Whatever classification is adopted it should never be more intricate than that necessary for actual use or more complicated than that which the forest sub-manager (i.e. the ranger or forester) can understand and use. The ranger or forester must not only be able to recognize the various types in the field but understand their value in silvicultural working so that he has confidence in using them.

GENERAL DISCUSSION

The above account has attempted to show how various features of a growing stock and the sites occupied interact and influence forest management which is constantly faced with choices between different activities to attain the best results. The forest manager, even when guided by a working plan, constantly has to make decisions—when to fell a tree or stand, how much to fell (finally or in thinnings), where to fell, what silvicultural methods to adopt, what species to use. From those decisions arise other problems which also affect the first decisions—the organization of men, materials and machines, sales, communications and so on. Good decisions cannot be made without good information. Conversely good information cannot be collected unless it is first decided what is wanted.

Guiding principles on the collection of information are:

(i) In even-aged forests the stand (or sub-compartment) is the unit of working. Each separate stand contributes to and influences all decisions of management. Therefore information must be based on the individual stand. In uneven-aged forests the working unit is the compartment, but the treatment unit is the individual tree as opposed to the stand.

(ii) Consequently inventories of the growing stock should give reliable information separately for each stand or other unit of working. That should also make it possible to analyse information for separate species, ages and quality of growth according to sites.

(iii) Records should also be kept separately for each stand or other unit of working to show its progress and outturn of material goods.

(iv) Owing to their nature stands cannot be permanent units. They should therefore be components of comparatively large territorial units (compartments or parcels) which can supply consolidated information for permanent parts of the forest. It is good if compartments or parcels coincide with or are whole components of utilization units or working blocks so that information can be consolidated for them too.

(v) Whole stands also form component units of felling series and working circles—but compartments may not. So information for them too can be consolidated from that collected for stands.

(vi) Mensuration techniques should conform with:

(a) Supply of information for each separate stand.
(b) Cheapness.
(c) Reliability and precision within chosen limits.
(d) Supply of information suitable both for planning silvicultural management and for satisfaction of markets.

(vii) The need for greater precision of information on volumes and increment increases with the age of stands or size of individual trees.

(viii) Owing to their greater cheapness sampling techniques are usually preferable to total enumerations and, since their precision is calculable, random sampling techniques are generally desirable. For older stands not only may a greater intensity of sampling be necessary but each individual stand may have to be sampled but for younger stands it may be enough to sample only the stratum to which each stand belongs and apply the general stratum results to each stand. Very small stands may all have to be assessed in that way. In uneven-aged forest large, permanent sample plots in each compartment might, for the sake of cheapness, replace the total enumerations of the method of control.

(ix) Stockmapping by ground and aerial survey (or even aerial survey alone with ground checks) is an essential preliminary to mensuration in order to map the forest and stand types which provide the stratification of population needed for mensuration.

(x) Local volume and yield tables are invaluable aids not only to the work of mensuration itself but for planning silvicultural management and forecasting the future performance and yields of stands. In uneven-aged forests the records from a series of repeated, comparable inventories of the growing stock and yields from it take the place of yield tables.

Bibliography

Aldridge, F., Hudson, R. H. (1955) 'Growing Quality Softwoods. A Critical Examination of the Turnbull Hypothesis', *Quarterly Journal of Forestry*, XLIX (4)

Anon. (1965) 'Forestry Commission; Working Plans Code'

Binns, W. O. (1966) 'Current fertilizer research', *Forestry Commission Supplement to Forestry*, 1966

Bruce, D. & Schumacher, B. S. (1942) *Forest Mensuration*. McGraw-Hill Book Coy, Inc, New York

Cajander, A. K. (1926) 'The theory of forest types', *Acta Forestalia Fernica* 31, Helsinki

Chapman, H. H. (1950) *Forest Management*, 2nd edition. The Hildreth Press, Bristol, Conn.

Conway, M. J. (1962) 'Aerial Application of Phosphatic Fertilizers to Radiata Pine in New Zealand', *Commonwealth Forestry Review* 41 (3), No. 102

Dawkins, H. C. (1958) 'The Management of Tropical High Forest with Special Reference to Uganda.' Commonwealth Forestry Institute, University of Oxford, Paper No. 34

Hewitt, E. J. (1966) 'A Physiological Approach to the Study of Forest Nutrition', *Forestry Commission Supplement to Forestry*, 1966

Hiley, W. E. (1955) 'Quality in Softwoods', *Quarterly Journal of Forestry*, XLIX (3)

Husch, B. (1963) *Forest Mensuration and Statistics*. The Ronald Press Coy, New York

Knuchel, H. (1953) *Planning and Control in the Managed Forest*, translated by M. L. Anderson. Oliver and Boyd, Edinburgh

Köstler, J. (1965) *Silviculture*, translated by M. L. Anderson. Oliver and Boyd, Edinburgh

Pardé, J. (1961) *Dendrometrie*. Imprimmé, Louis-Jean-Cap

Phillips, E. W. J. (1962) 'The Anatomy of Softwoods and its Influence on Timber Quality', *Forestry Commission Supplement to Forestry*, 1962

Rendle, B. J., Phillips, E. W. J. (1958) 'The Effect of Rate of Growth (Ring Width) on the Density of Soft-woods', *Forestry* XXXI (2)

Rennie, P. J. (1957) 'The Uptake of Nutrients by Timber Forest and its Importance to Timber Production in Britain', *Quarterly Journal of Forestry*, LI (2).

Rutter, A. J. (1966) 'Water Relations and Growth in Trees', *Supplement to Forestry*, 1966

Spurr, S. H. (1952) *Forest Inventory*. The Ronald Press Coy, New York

Spurr, S. H. (1960) *Photo-grammetry and Photo-interpretation*. The Ronald Press Coy, New York

Chapter VIII

THE YIELD AND ITS
REGULATION

This chapter will be devoted to the yield of wood and regulation of that yield. It will not consider the yields of other material goods or intangible services except in so far as they may influence the yields of wood. Let us first examine the motives for cutting trees for a yield.

MOTIVES FOR CUTTING

The management of a forest for the production and supply of wood obviously requires the continuous annual (or occasionally periodic) cutting of individual trees or stands of trees. But it is not so clear that there are several interconnected motives for cutting which affect the type of cutting and its regulation in quantity, quality, place and time. Those motives, which are all affected by the objects of management, are:

(i) *The supply of wood to consumers*

For this prime purpose of cutting the forest must be so worked that there are available the species, sizes, qualities and quantities of wood that suit the chosen or available markets. Also the trees must be reasonably concentrated in space to make their felling and collection worth while. For example, the harvesting of small scattered supplies of even a valuable species of timber may be too expensive to suit a market. Another critical factor may be the need to supply sufficient and regular quantities of timber (perhaps to a close specification) that are essential for the support of industries and markets which may be expanding. There is then a danger that either fellings will be inadequate to meet demands or fellings will be made in excess of the capacity of the forest so that an industry may eat up and destroy first the forest and then itself in consequence.

Therefore, for the continuous management of a forest, as opposed to its exploitation and destruction, supplies to markets depend on motives (ii), (iii) and (iv) below.

(ii) *The tending of the growing stock to maintain or improve its capacity to produce the desired goods in their best form and quality as quickly as possible*

The tending concerned is mainly thinnings whose influence is fourfold:

(*a*) Control of the specific composition of the crop by removal or retention of species according to their values either as a material product or as a silvicultural aid to other more materially valuable species. For example the presence of a deep-rooted, low value species among a shallow rooted and more commercially valuable species may help in promoting soil fertility and the increment and yield of the more valuable species.

(*b*) Removal of inferior stems to give space to better stems and thereby concentrate growth on them.

(*c*) Control of the diameter increment of individual trees so that the rate of growth is compatible with the quality of timber desired and marketable sizes are reached earlier and rotation length reduced accordingly.

(*d*) Supply of early net income.

If all thinnings are saleable at a profit there is no hindrance to the full use of those influences. If thinnings are not saleable at a profit careful judgement is necessary to decide whether the cost of the thinning is balanced by the ultimate benefits that they produce.

Another type of tending felling is regeneration felling whereby final fellings are controlled and made over several years to isolate mother trees in the numbers and for the length of time most suitable for natural regeneration. The intensity and frequency of such fellings must vary with the speed and success of the regeneration and therefore also with the weather, site and species. So the yield from them will tend to fluctuate from year to year.

(iii) *The adjustment of the quantity and form of the growing stock to that most suited to the objects of management*

This purpose or motive of cutting concerns the adjustment of the age-classes (or size-classes in an irregular forest) and there are three main objects of management which affect the need to adjust age-classes, namely:

(a) *A sustained yield.* To attain a sustained yield the final fellings must be so made that the distribution of age-classes is brought steadily closer to the normal distribution. To do so will inevitably necessitate from time to time the felling of some stands before or after they are mature.

(b) *A sustainable yield.* A sustainable yield is one that never falls, so temporary increases in yield are not admissible but only those that can be permanently maintained. A sustainable yield may be essential for an industry which is dependent on a particular forest for its raw material. The growth of such an industry will then also depend on a rising sustainable yield and accurate forecasts of it.

For a forest to provide a sustainable yield it is almost certain that over-mature stands will always have to be husbanded to compensate for fluctuations in the sequence of maturing age-classes and for accidental wind or other damage. The final goal is a sustained yield from a normal forest.

(c) *A sustained production.* For sustained production any stand which is failing to grow at a satisfactory rate should be felled and replaced. Stands are felled when they reach the peak of their performance or when their replacement by another species will give a better result.

Sustained production conflicts with the attainment of both sustained and sustainable yields since neither can over-mature stands be kept and husbanded nor can under-mature stands be felled earlier to compensate for an expected shortage of mature stands. So, as the emphasis is on production, unthrifty stands are felled and replaced whatever their age so that the distribution of age-classes and yield are likely to fluctuate. Those fluctuations will be the most pronounced in small forests or felling series where there is not enough scope for variety in species, sites, age-classes and accidents to balance each other. In a region of several forests the average annual regional yield may be comparatively uniform; yield fluctuations in the several forests will tend to balance each other.

(iv) *Position of fellings*

It may be desirable to cut stands out of turn to provide protection for other stands (from wind or sun) or to arrange a better juxtaposition of age-classes for convenience of utilization or silvicultural advantage. Necessity of this kind is particularly evident when introducing a special silvicultural system—such as a strip or wedge system—and in windy, hilly tracts where belts or groups or trees must be kept to prevent wind damage to other stands.

Those purposes or motives for cutting, interconnected and possibly to some extent in conflict, all aim to supply the material goods wanted and at the same time to improve the capacity of the forest to produce them. Therefore they provide the basis on which the forest manager decides his cutting policy—namely how much to cut, and where, when and how to cut it. Together those decisions constitute the regulation of the yield.

TYPES OF YIELD

In even-aged forestry there are two types of yield—the final or main yield and the intermediate yield from thinnings. In general one can say that the final yield, i.e. the removal of stands and their subsequent regeneration, controls the proportional distribution of age-classes whereas the intermediate yield from thinnings controls the economy by providing early returns, influencing the length of rotations and controlling timber quality.

In irregular forestry the final and intermediate yields are indistinguishable and combined in one felling operation; naturally they have a combined effect on size-class distribution, the economy and exploitable age. But owing to the indivisibility of the two types of yield a different approach to yield regulation may be needed from that used in even-aged forestry.

A matter which is always relevant is the greatest productive capacity of the forest, i.e. the greatest quantity of wood that can be cut continuously. It is true, however, that the eventual productive capacity of a forest may not be approachable for so long a time that it is incalculable with any precision so that only the yield in preliminary phases has much importance. Even then some consideration of the probable eventual production of the forest has relevance to current planning, e.g. for the introduction of industry later.

Whether fluctuating yields are acceptable or not the maximum *average* annual production of a forest over a long period cannot exceed that of the current annual increment of a normal forest, i.e. that forest which has a normal distribution of age-classes for the species and rotations considered the best for the available sites and objects of management. If an abnormal distribution is maintained to avoid the sacrifices necessary to attain normality there will be continued fluctuations in the yield commensurate with the fluctuations in the age-classes and the average age of the forest. But the *average* yearly yield will not exceed the increment and yield of that forest when normal although some sacrifices (e.g. husbanding old stands with a lower increment while young stands are maturing), and loss of increment will have been avoided during a period of conversion to normality. But the maximum average annual output cannot exceed the normal sustained yield.

To understand methods of yield regulation it is, therefore, necessary to consider first how the yield is regulated for growing stocks that are managed for sustained yields as the prime object. That has been the classic attitude to regulation of the yield and has influenced forestry since the end of the eighteenth century. Some consideration will then be given to situations when a sustained yield is not essential.

METHODS OF YIELD REGULATION

I. SUSTAINED YIELDS PARAMOUNT

General requirements and methods

Regulation of the yield requires firstly a calculation of what the amount of the yield should be, secondly an apportionment of that yield to thinnings and final fellings and thirdly, the construction of a cutting plan which determines the identity of stands to be felled or thinned and the time of their felling or thinning. How precise or rigid each of those elements of yield regulation should be will depend on circumstances. For example, the cutting plan may rigidly prescribe the very years in which particular stands should be felled, or it may flexibly explain only the principles that the manager should observe in choosing the annual coupe or felling area himself. It is also desirable that there should be a control or annual comparison of what has been prescribed in the plan of work with what is actually done, supported by an explanation of differences. Unless there is such a control, undesirable divergencies between planning and action are liable to occur and multiply.

Methods of calculating the desirable yield naturally depend on the constitution of the growing stock, especially its structure and silvicultural organization, and on the extent of knowledge of its volumes, increments, rotations used and of the state of individual stands. The many methods of yield calculation can be classified by the variables used as shown below:

1. *Area*
   (a) Control by silvicultural or other felling rules.
   (b) Control by rotation and age-classes or periodic blocks.
   (c) Control by development or treatment classes.
2. *Volume* Control by rotation or exploitable age.
3. *Volume and Increment* Control by rotation or exploitable age.
4. *Numbers of Trees* Control by stem size and increment (time of passage).

Let us consider each of these main methods separately.

A. Area methods

Definition of yield in basic terms of area is very common and is found in several forms all of which involve areas of forest that are felled or tended in a given time. The more developed and refined methods may, as in the several periodic block methods, prescribe the yield by the volume of timber to be cut but none the less the basis of the method remains the removal and regeneration of the growing stock on a definite area within a definite time as the volume yield prescribed is merely that which is found on a certain area in the time prescribed.

(a) Control by silvicultural or other felling rules

This very simple method is often used in extensively managed tropical forests particularly in those which are more or less irregular and under an early phase of management. A felling cycle is fixed and the forest divided into an equivalent number of sections and one section is worked in turn each year. The quantities of trees felled each year is controlled only by felling rules which use simple silvicultural principles such as species to favour, removal of mature, over-mature and diseased stems, retention of sufficient seed-bearers, the freeing of established regeneration, saplings and poles. Minimum size-limits, by stem diameter or girth, may also be used to indicate acceptable maturity and marketability of trees.

Provided that the felling rules ensure that useless species and diseased and over-mature stems are removed or killed and provided that regeneration establishes itself, the method encourages an improvement of the growing stock and initiates a commercial use of the forest. But there is little or no control of the quantities of timber felled annually so that yields will fluctuate unless the sizes of section worked each year are varied to make then equi-productive. The next stage of management, after collection of data on rates of growth and numbers of trees in the various size-classes have been collected, would be to control yields by numbers or basal area of trees felled.

Forecasts of out-turn can and should be made, however. This can always be done at the time of marking the trees to be felled but longer forecasts can also be made by sample enumerations and sample markings in sections to be worked several years ahead.

The method can also be applied to thinnings in more intensively managed even-aged forests by fixing a thinning cycle and allotting stands accordingly, to be thinned in turn according to thinning rules which may or may not be expressed quantitively, such as by basal area.

(b) Control by rotation and age-class distribution

The final yields of even-aged forests can be very satisfactorily regulated by the simple relation between forest area and rotation. Thus in a normal forest the annual final yield would equal the quantity of trees growing on an area which equals the total forest area divided by the rotation—i.e. expressed by area the annual final yield, AY, $= A/R$. By using equiproductive areas the volumes of trees felled yearly would be stabilized.

The intermediate yield from thinnings could also be regulated by area by fixing a felling cycle and thinning equal areas of similar age-classes each year in turn as explained in the previous section.

The same formula for the final yield, $AY = A/R$, could be used in an abnormal forest and a normal series of age gradations would then be attained in one rotation if the programme of felling and regeneration were fulfilled and all were to go well. It is the quickest and simplest way to achieve a sustained yield. But sacrifices would be increased and they would be heavy in a very abnormal forest as either immature trees would be felled before they were ripe or the felling or mature trees would be delayed until they became over-mature, lose increment or even die.

It is usually advisable to compromise. If there is an excess of mature timber more than the normal area should be felled for a time; if there is a deficit of mature timber less than the normal area should be felled until the excess or deficit is largely corrected. Constant reappraisals of the state of the age-classes and of the yield to be cut, e.g. at ten-year intervals, will steadily bring the forest to normality.

A useful basis for estimating the compromise is to combine the areas of the oldest two or three age-classes and then calculate the yearly yield by which those age-classes are removed in the time equal to the sum of their combined class intervals. For that purpose the stands which are overmature should be included in the mature age-class. The process is shown in the simple example in Table 12, for a felling series whose area is 120 acres, the rotation is sixty years and age-classes are in ten-year intervals.

In that example three features are noticeable. Firstly, there is a fairly large fluctuation in yield, secondly over-mature stands continue to exist for over twenty years, and thirdly immature stands will begin to be cut in the third ten-year period. Decision on the actual yield to be cut will depend on the state of the over-mature stands and the importance of a steady yield. Thus, if the over-mature stands are not too old and poor, the cut in the first period might be reduced to 26 acres (2·6 yearly) and changed to 22 acres in the second period at the end of which the age-class distribution would then be 22, 26, 10, 15, 30, 5, and 13 acres of over mature stands. It might then be possible to introduce the normal cut of 20 acres in a period without undue sacrifice. If greater importance is attached to removing old and unproductive stands then the over-mature stands must be quickly removed and regenerated thereby perpetuating to a large degree the poor balance of age-classes and a fluctuating yield.

The example enables one to realize the conflict that nearly always arises between various advantages and disadvantages. Here the conflicting desires are an eventual sustained yield, avoidance of present fluctuations in yield and reduction of losses that result from growing unproductive stands.

Table 12

FOREST AREA BY AGE-CLASSES WITH ANNUAL FINAL YIELD CALCULATION (ACRES)

every 10th year for 30 years

Age Class Years	Nor-mal Areas	Actual Areas Now	Annual Yield Calcn. now	Areas after 10 Years	Annual Yield Calcn. in 10th Year	Areas after 20 Years	Annual Yield Calcn. in 20th Year	Areas after 30 Years	Annual Yield Calcn. in 30th Year
0–10	20	10		30		18		24	
11–20	20	15		10		30		18	
21–30	20	30		15		10		30	
31–40	20	5		30		15		10	
41–50	20	10		5		30		15	
51–60	20	15	$\frac{60}{20}=3\cdot0$	10	$\frac{35}{20}=1\cdot8$	5	$\frac{47}{20}=2\cdot4$	23	$\frac{38}{20}=1\cdot9$
61 and over	—	35		20		12		—	
TOTAL	120	120		120		120		120	

Calculation by formula. The annual area yield may also be calculated by formula if one assumes that the ratio of the annual yield (AY) to the actual average age (AA) of the forest should equal the ratio of the normal yield (NY) to the normal average age (NA), i.e.

$$\frac{AY}{AA} = \frac{NY}{NA} \quad \text{and} \quad AY = \frac{AA \cdot NY}{NA}$$

Then in our example, if the over-mature stands are aged sixty to seventy years, the average age of the forest may be taken to be

$$\frac{(10\times5)+(15\times15)+(30\times25)+(5\times35)+(10\times45)+(15\times55)+(35\times65)}{120}$$

= 39·7 or say forty years.

(i.e. the sum of the products of the area and mean age of stands divided by their total area).

Now the normal average age of the forest is thirty years and the normal annual yield = 2 acres.

So the calculated annual yield of the forest, AY,

$$= \frac{\text{AA . NY}}{\text{NA}} = \frac{40 \times 2}{30} = 2 \cdot 7 \text{ acres}$$

That $2 \cdot 7$ compares with the $3 \cdot 0$ acres of the previous calculation, and gives an excellent result.

But it must be remembered that the formula (like most others) can often give absurd results. Thus the forest of our example might contain stands *only* thirty years old and so have an average age equal to the normal and yet not one mature tree would be available. Many other abnormal distributions would also give an average age equal or nearly equal to the normal, so that the formula must be used with great caution.

Periodic yields and a regeneration block. As suggested in the above example the yield may be regulated by areas to be felled during a period rather than annually so that there can be some flexibility in amounts cut in any one year during a period. If that were to be done the area to be felled in the period could be selected as a regeneration block. Then the standing volumes of the stands in the block could be measured and also their increment. By calculation the area yield could be connected to a volume yield equivalent to that area, namely a volume yield which, if cut regularly, would result in all the trees in the area being felled by the end of the period.

If the trees were not growing then that yearly volume yield would be their present volume, V, divided by the period, P. But they do grow. Now, the trees cut at the beginning of the period have no time to grow but the trees cut at the end of the period can grow for the whole period. So, on average, the trees put on half the increment (I) they would have put on if they had been left untouched. So the volume available for cutting annually in the regeneration block is the present volume (V)+half the increment (I) that the untouched stands would have put on in the period.

Expressed in symbols, known as *Cotta's formula:*

$$\text{AY} = \frac{\text{V} + \frac{1}{2}\text{I}}{\text{P}}$$

or

$$\text{AY} = \frac{\text{V}}{\text{P}} + \frac{1}{2}\text{i}$$

(where i = the *annual* increment of V).

Conclusion. So one can recognize three phases in regulating the yield by area and rotation—an annual yield by area, a periodic yield by area and a volume yield equivalent to a periodic area yield—three phases of increasing flexibility.

The first two phases suit clear felling silviculture only. The third phase, however, also suits an even-aged silviculture that uses successive fellings for natural regeneration.

Thus, for natural regeneration under the Uniform System of silviculture a suitable block of stands is chosen for regeneration within a period, say twenty years. The yield is then fixed by volume by Cotta's formula. It does not then matter where or when the regeneration fellings are made in the block or how heavy they are per acre, provided that they are completed in the period within the selected block and conform to the volume yield. So the felling can be varied in intensity and position to suit the progress of the regeneration.

The convenience of adapting what is basically an area method of yield regulation to an equivalent volume yield encouraged the development of the periodic block methods of silvicultural management. These we shall now consider.

Periodic block methods of management

The development of forest management in Europe in the early years of the nineteenth century led to an emphasis on orderliness as well as crowded stands lightly thinned. The attainment of the normal forest and sustained yields was also greatly emphasized at a time when transport was slow and laborious and when changes in the types of wood wanted were comparatively small although the amounts of demand were steadily increasing. So it was generally accepted that appreciable sacrifices were justified to attain normality. Moreover natural regeneration was generally considered to be desirable.

Previous working, particularly that of hardwood high forests in France, had relied on generally successful methods, such as the French 'Tire et aire' method by which area coupes with natural regeneration had preceeded more or less regularly and consecutively round the forest during a long rotation. To modernize the old system and lead it to the new normality a practical and logical method was to organize age-classes in a series of blocks or groups of compartments whereby different operations would be concentrated in separated blocks. Thus regeneration fellings would be in a block of mature age-classes, weeding and cleaning in an establishment block of young age-classes, thinnings in the rest sub-classified as necessary according to the average age of the intermediate blocks.

Since the mid nineteenth century periodic block methods have been modified and developed but are still used especially in France. They may be classed in three types—the permanent, the revocable and single periodic block methods.

(1) *The permanent or fixed periodic block method.* This method was the first to be adopted. It is rigid in design, in both space and time, in order to attain a very orderly, simple and convenient pattern which segregates the main operations of final regeneration fellings, tending young crops and of thinnings in compact permanently defined blocks. Inevitable initial sacrifices are accepted to be worth the eventual advantages.

The method applies three basic principles—firstly the use of one rotation only throughout the felling series; secondly the selection of a management and regeneration period which must be a simple factor of the rotation in order to define the number of periodic blocks which equals the rotation divided by the period, e.g. $R/P = 100/20 = 5$; and thirdly, the periodic blocks must have equal areas (or possibly equi-productive areas) and form compact groups of contiguous compartments.

The one rotation adopted will be that which on average produces the mean exploitable diameter wanted. If there are more than one species the rotation must be the one that suits the main species only or the one which is the best compromise of the several best specific rotations—which is clearly unsatisfactory but essential to the method.

The period selected should be short enough both to ensure regularity of stands after the regeneration of a block and to prevent management prescriptions for treatment of each block from becoming out of date and inapplicable. In France a period is usually between twenty and thirty years.

Consequently each compact, self-contained block will contain its requisite group of age-classes. Thus, if the rotation is 100 and the period is twenty years there will be five P.B.s; P.B. I will be under final, regeneration fellings and will ideally contain age-classes 81–100, P.B. II classes 61–80 and so on. At the end of the period P.B. II becomes the final felling and regeneration block in its turn. The initial allotment attains as close an approximation to the ideal as is allowed by the existing distribution of age-classes and the necessity for blocks to be equal in area, compact and self-contained. Some incongruity of age-classes in their respective blocks is certain to occur in the initial allotment.

The periodic blocks are then demarcated on the ground and so remain permanently as first selected. Being felled and regenerated in strict sequence by periods, P.B. I in the first, P.B. II in the second period and so on, the felling series will be normal at the end of the first rotation and each P.B. would be suited to distinct types of treatment and out-turn or yield, and each type concentrated in compact areas on the ground.

Yield regulation is simple. The final yield is by volume calculated by Cotta's formula for the final felling or regeneration block. The intermediate or thinning yield is also regulated basically by area according to a thinning cycle which can differ in length from block to block in order to suit the average age of the stands in the block. Forecasts of thinning yields can be based on past experience or on yield tables.

It is notable that the final yield is prescribed as a definite annual volume yield. But for natural regeneration it is essential both to take advantage of good seed years and to adapt fellings to the progress of regeneration. Those needs demand flexibility both in the site of felling in the regeneration block and in the quantities felled in any one year. Consequently it is usual to allow large divergencies from the annual yield provided that the yield prescribed for the period is closely observed. Thus in France up to five times the prescribed annual yield may be cut in any one year provided that the amount cut does not exceed the volume left standing in the block (Anon. (1964), Manuel pratique d'aménagement).

The fixed periodic block method has severe drawbacks:

(i) Inevitable sacrifices arise from the initial selection of the compact blocks which will certainly contain stands incongruous with the proper age-class.

(ii) The rigidity is too great. Even if regeneration is attained punctually accidents may happen elsewhere. Wind, insect or other damage outside the regeneration block may make regeneration there advisable, out of turn. Then either the method has to be modified or abandoned or the regeneration is delayed and loss of production results.

(iii) The one rotation can hardly be varied to suit varying growth rates on different sites or of several species or varying markets. So the method is only suitable for an area which is uniform in site and for one main species which is both at home and suits one main object of management requiring one rotation.

But the considerable advantages of simplicity and orderliness, compact location of different treatments (and sizes of timber) and achievement of normality in one rotation have given the method a great vogue. But, except in one or two forests in France where sites are uniform, regeneration is not too difficult where the species is happy and healthy and a single rotation satisfies the objects of management, the method has now been abandoned.

(2) *The revocable periodic block method.* To avoid the rigidity and reduce the sacrifices of the permanent P.B. method the obvious remedy is to abandon fixed P.B.s and re-allot them according to circumstances at each revision of the working plan, namely at the end of each period. As parts of a P.B. will perform differently from other parts, a re-allotment will result in the fragmentation and dispersion of P.B.s so that they are no longer compact and self-contained even if they were compact at the first allotment. The unit of management, in effect, has now become the compartment, or even possibly the sub-compartment (stand), and not the P.B. although prescriptions of the current plan will apply to compartments grouped in P.B.s.

The re-allotment of compartments to the several periodic blocks (which need no longer be compact) at the end of each period has far reaching consequences. No longer must the whole of an originally selected P.B. be regenerated at the end of a single, definite rotation. On the contrary a compartment which develops quickly can be re-placed in the regeneration block before the end of the average rotation and one which develops slowly afterwards. So individual compartments can have different rotations according to their specific composition and site quality. If the sub-compartment or stand is made the unit of working instead of the compartment then each stand can have a separate rotation. That greater flexibility could be increased even more if the period is shortened so that re-allotment to periodic blocks is made more frequently. But the greater fragmentation that would result might make the method unworkable.

The loss of compactness of blocks is an undoubted disadvantage for organizing silvicultural and utilization works. But the ability to treat compartments (or even stands) according to their performance should greatly increase their productivity. Final and intermediate yields are calculated in the same way as in the fixed P.B. method. But the areas of blocks (and they should all be equal) is affected by the several rotations that may be in use. Thus:

$$\text{Area of P.B.} = P\frac{A_1}{R_1} + \frac{A_2}{R_2} + \ldots \frac{An}{Rn}$$

where $P =$ the period, A_1, $A_2 \ldots An =$ the areas of the numerous species and sites requiring different rotations and R_1, $R_2 \ldots Rn$ are their respective rotations. This complication is necessary in order to equalize yields by relating areas in each P.B. (i.e. areas of age-classes) to the rotation, which will be the mean of the several rotations used weighted by the area subjected to each individual rotation.

(3) *The single periodic block method.* In the third periodic block method only the regeneration block is allotted and, similarly to the revocable P.B. method, the compartments comprising it need not be contiguous. At the end of the period when regeneration should be complete, or at revision of the plan a fresh allotment is made. Therefore the regeneration block moves or floats round the forest so that the method is sometimes known as the floating periodic block method.

The remaining compartments are usually placed in two groups. The one group is composed of those, not necessarily contiguous, compartments that will probably be placed in the regeneration block in the next period. They will be subjected to the types of thinning and other treatment designed to prepare the stands for regeneration. The other group is composed of the younger compartments—saplings, poles and middle aged stands under cleanings and thinnings. If desired the younger stands subjected to weedlings and cleanings would be separated from those being thinned.

The mean rotation, or age of exploitability of the average sized tree desired, is fixed in the same way as in the revocable periodic block method. As there is freedom in selecting compartments (or even stands) for allotment to the single (regeneration) block at the end of each period or at plan revision it is clear that the single P.B. method can be used for a felling series composed of even-aged stands that have different rotations—for reason of species or site. Also, as in the revocable block method, the area of the single, regeneration block is:

$$\text{Area} = \text{Period} \frac{A_1}{R_1} + \frac{A_2}{R_2} + \cdots \frac{An}{Rn}$$

As the area of the single block is related to the period and the mean age of exploitability and must be felled and regenerated in the period, continuance of the programme in subsequent periods will lead the forest towards normality.

Choice of the single, regeneration block. The selection of compartments (or, if desired, sub-compartments) for regeneration will depend on the actual state of the growing stock—age, health vigour, density, etc.—and the consequent urgency for felling and regeneration in the period. In their choice, therefore, first consideration can be given to insect or wind damage and the growth performance of any compartment (or sub-compartment) with respect to the objects of management.

Now in a forest of area A whose average rotation is R and the selected period is P, the area of the single periodic block, AB, $= (P/R) . A$. But the area of compartments or stands which it is considered to be *desirable* to allot to the single block for felling and regenerating in the period will almost certainly differ from and be greater or less than the area AB, unless the area of mature stands happens to correspond with the normal area. Then if the exact *desirable* area is allotted to the single block any imbalance of age-classes will be continued.

Unlike the permanent block method with its single and equal compact blocks and the revocable method with its blocks dispersed but equal in area which thereby compel progress towards sustained yields and normality, the single block method is extensively flexible since the area of the single block may be anything (Anon. 1964) and the normal gradation of age-classes is not necessarily implied. Owing to the emphasis on one periodic block the distribution of age-classes can be overlooked and become or remain abnormal.

In selecting the area of the single block, therefore, it is necessary to consider not only the state of individual stands or compartments but also the capacity of the forest as a whole to fulfil the objects of management. If the only desire is to obtain maximum productivity and if uniformity of yield and the concomitant uniformity of regeneration and other work are not important, then selection of the single block can be made solely on the actual conditions and performance of each stand or compartment. But if sustained yields and sustained working are important, the area of the single block must be related to the general state of the forest and its age-class distribution. The forest as a whole can then be led towards sustained yields and working.

There are several methods of approach to selection of the area of the single block. To simplify explanation let us consider a simple example, namely a forest of 200 acres whose rotation is 100 years and the chosen period is 20 years. Then by the formula $AB = P.A./R$ the area of the single block should $= 20/100 \cdot 200 = 40$ acres, and it should be felled and regenerated in 20 years.

Let us suppose, however, that the forest contains 60 acres of over-mature, mature and open stands of which it is urgent to fell and regenerate 25 acres as quickly as possible and the remaining 35 acres within the 20 years if possible.

Action then can be taken in several ways:

(i) Accept the standard formula and fell only 40 acres in the period by prescribing an annual volume yield of $V/20+\frac{1}{2}i$, where $V =$ volume on 40 selected acres and i is the annual increment of V.

(ii) Fell all the 60 acres but extend the period of removal in proportion to the excess felling thus:

$$AB = \frac{P}{R} \cdot A \qquad P = \frac{ABR}{A}$$

So, if $AB = 60$ acres, $P = (60 \times 100)/200 = 30$ years. Then the annual volume yield $= V/30+\frac{1}{2}i$, where V and i are the volume and annual increment on the 60 acres.

It is clear that this method enables the 25 acres to be felled first but the rate of felling is equivalent to the 'normal' rate. At the end of the standard period of 20 years there will still be 20 acres of the block unfelled; the position is then re-assessed.

(iii) Fell the 60 selected acres in the standard period of 20 years—and perpetuate the mal-distribution of age-classes, so that the volume yield $= V/20+\frac{1}{2}i$ where V and i are the volume and annual increment on 60 acres.

(iv) Adjust the 60 acres first shown for regeneration to a figure based on a study of the forest and its age-classes as a whole and of the forecasted progress of age-classes to maturity (see method described earlier under A(*b*), control by rotation and age-classes). The figure so obtained might be 50 acres for regeneration in the period of 20 years.

Then the volume yield $= V/20 + \frac{1}{2}i$ where V and i are the volume and annual increment on 50 acres which are finally chosen in order of priority by urgency for felling according to the state of individual stands or compartments.

It can be seen that method (i) and (ii) are similar in principle and stress only the improvement of the age-class distribution, method (iii) stresses only improvement of production at the expense of age-class distribution and (iv) is a compromise based on forecasts of the development of age-classes, a compromise which might have been again altered to, say, 55 acres if productivity of stands were to be given greater weight than age-classes.

If method (iv) is adopted it is clearly necessary to decide very carefully what the relative values of several influencing factors or constraints are and to compare the results of different courses of action. To achieve the best compromise the use of a computor and linear programmes may be desirable so that the results of alternative courses of action and the respective influence of the various constraints can be compared and assessed.

Subsidiary principles of application of the method. The area allotted to the single block should be completely felled and regenerated in the period. But particularly when natural regeneration is used the regeneration may not be completed exactly to programme; a portion may still contain scattered mother trees and the process of regeneration must be finished in the next period.

To allot the whole surface area of the partially regenerated portion to the next period is clearly wrong. So the acreage carried forward is calculated as that proportion of its superficial area which the tree volume (or basal area) bears to the volume (or basal area) of untouched growing stock. Thus if the volume or basal area per acre of the remaining mother trees is one quarter of that of an analogous but untouched stand, then three quarters of the superficial area occupied by the mother trees is deemed to have been regenerated and one quarter is carried forward for regeneration in the next period.

It should be noted that such a carry-forward reduces the area of the new block to be selected for the next period by the same amount (if the formula $AB = P/R$. A is observed). In effect the rotation is being prolonged accordingly. The best procedure would be to ignore the *area* partially regenerated and credit the volume and increment on it to the area of the next block so that the sum of the two volumes is removed in the next period. No delay then results.

That is all very well when the area to be carried forward is small and unimportant. Only a slight variation from the proper rate of regeneration and a slight disorder in the distribution of age-classes result. But, if the area is substantial and if the failure to complete the programme of regeneration continues in succeeding periods, the implication is that the single periodic method is failing. Either the regeneration techniques must be improved (e.g. by supplementing or displacing natural with artificial regeneration) or another method such as the regeneration area method instead of the single periodic block method should be used.

To adhere to the regeneration programme in the allotted time is clearly an essential element in the single periodic block method. Therefore to ensure the planned progress of fellings and regeneration it is also essential that the volume yield prescribed should correspond accurately with the volumes and increments of the stands in the single block. Checks on the progress of fellings should be made from time to time during each period and the yield adjusted as necessary to ensure punctual completion.

Similarly delays resulting from failure to obtain natural regeneration punctually must be avoided by, for example, supplementary artificial planting, improved ground preparation or weed control.

General comments on the single block method. The essential points of the potentially very flexible and also practical single block method are summarized below:

(i) The area selected for the single block has to be felled and regenerated during a definite time (the period). This factor distinguishes the single periodic block method from the regeneration area method which will be described later.

(ii) The final and intermediate yields are assessed and prescribed separately; the final yield (from the single block) is prescribed by the volume equivalent to the area of the block; the intermediate yield (from the rest of the forest) is usually assessed and prescribed by area—with forecasts of outturn if necessary. This factor also distinguishes the single block method from the regeneration area method which requires a volume yield which embraces both final and intermediate yields.

(iii) The completely fresh allotment of the single block after each period and the absence of any restraint on the dispersal of stands or compartments allotted to it enables more than one rotation to be used and does not limit removal ages. Even-aged stands of different species and on various sites can therefore be accommodated by the method as in the revocable block method. But it is necessary to be able to measure the area (and volume) of stands with reasonable ease and close accuracy. So if stands or sub-compartments rather than whole compartments are the units of working it is essential for them to be fairly large and not mere even-aged clusters of trees. If they *are* small another system suited to a more irregular growing stock becomes necessary.

(iv) If it is accepted that the extent of the single block can differ from $AB = P \cdot A/R$ and that a controlled or normal distribution of age-classes is unimportant, the method can be used with infinite flexibility to suit the state of the stands relative to the objects of management.

(c) Control by development or treatment classes

Consideration of the periodic block methods shows that the management of even-aged forests requires not only the regulation of quantities cut each year but also their location in compartments or sub-compartments (whichever is the unit of working) to facilitate and intensify silvicultural tending and utilization. Only then can treatment appropriate to the actual state and performance of stands be efficiently devised and the distribution of age-classes be controlled to suit the objects of management.

Those needs were recognized long ago. In Germany Judeich's

Bestandwirtschaft or the stand management method was developed from 1811, chiefly by Cotta and then by Hufnagl to be finally made more precise by Judeich in about 1870 (Schlich, 1925).

In short Judeich assigned woods (i.e. compartments or sub-compartments) to the silvicultural treatment best suited to them to attain full, healthy increment and to final felling when they reached financial ripeness. The assignment would take into account the desirable location of age-classes (for protection, utilization and general organization) and also, if sustained yields were required, the proportional distribution of age-classes. The treatment and prescription would be for a short time only (ten years) before revision and the yield would be the sum of the volume outturn from the treatment allotted to each wood.

Judeich's method, therefore, is similar to the French single block method but employs a more detailed allotment of younger woods to a particular treatment and arrangement and also more frequent revisions. The crux of both methods is the quantity to be allotted to final fellings if a sustained yield is the basic object. Judeich used a 'modifying regulator' based on the area of the forest and the rotation in the manner suggested in paragraphs under section 1(*b*) above (control by rotation and age-class distribution).

Development and treatment classes

If variable site conditions and use of several species result in different growth rates and require the use of several rotations and variable removal ages, a classification of stands by age-class alone gives a confusing impression of the sequence of maturing stands. Thus, a forest might contain an equal distribution of age-classes up to fifty years. But if half the forest were to contain Douglas fir maturing at fifty years and the other half Scots pine maturing at eighty years, one half would be normal and the other half far from normal. The combined age-class table gives a very erroneous impression of the forest as a whole.

If stands are classed according to their development stages which can be related to treatment a very different impression is given. A slow growing species would pass through a development or treatment class slower than a fast growing one so that the classification would give at least an approximate representation of the degree of maturity of each stand and the forest as a whole. That would be particularly useful when the ages of stands have not been accurately recorded and yield tables are unreliable or non-existent.

A possible classification with approximate equivalent age-classes for two sample species is given in Table 13.

Table 13

DEVELOPMENT AND TREATMENT CLASSES

Class No.	Development Class	Treatment Class	Relative Maturity %	Equivalent ages for 2 species	
				Rotn. 50 yrs.	Rotn. 75 yrs.
I	Establishment	Weeding, cleaning	0–20	0–10	0–15
II	Small pole	Early thinning	21–40	11–20	16–30
III	Large pole	Middle thinning	41–60	21–30	31–45
IV	Young timber	Late thinning	61–80	31–40	46–60
V	Mature timber	Final felling	81–100	41–50	61–75
VI	Over mature	Salvage felling	>100	>50	>75

Yield calculation. The final area yield, for the object of leading the forest towards a sustained yield and normality, can be calculated as a percentage of the forest area in the same way as in section A(*b*) page 172, using percentage instead of absolute areas. To do this a time factor must be introduced, namely the average or guiding rotation.

Let us consider a simple example for which the basic data are:

Area of forest—300 acres

Tree species—A on 200 acres with a rotation of 50 years.

B on 100 acres with a rotation of 75 years.

$$\text{Average rotation} = \frac{(50 \times 200) + (75 \times 100)}{300}$$

$$= 55 \text{ years.}$$

Let the actual and normal areas in each treatment class be as shown in Table 14 below, areas in acres and percentages:

Then taking the forest as a whole and using the method of

section A(b), it may be considered desirable to remove classes IV, V and VI in two-fifths of the average rotation—i.e. 52 per cent of the forest area in two-fifths of fifty-five years = twenty-two years.

$$\text{Annual Yield} = \text{AY} = \frac{52}{100} \times \frac{300}{22} = 7 \cdot 1 \text{ acres.}$$

So if yields and prescriptions are revised decennially the yield would be fixed at 71 acres over ten years. It could then be left to the discretion of the manager which actual stands should be felled in either species A or B in any one year.

Table 14

Treatment	Species A		Species B		Whole Forest	
	Normal area	Actual area %	Normal area	Actual area %	Normal area	Actual area %
I	40	50 25	20	30 30	60	80 27
II	40	30 15	20	10 10	60	40 13
III	40	15 7	20	10 10	60	25 8
IV	40	15 8	20	20 20	60	35 12 ⎫
V	40	60 30	20	25 25	60	85 28 ⎬ 52%
VI	—	30 15	—	5 5	—	35 12 ⎭
TOTAL	200	200 100	100	100 100	300	300 100

The yield could be worked out separately for each species in the same way if desired.

Three things may be noted in this flexible method in which measurements of increment and volume can help subjective assessments of the class to which a stand belongs.

(i) No stand is placed in its class owing to its age. Thus any stand, owing to site conditions, may develop quicker than its neighbours. Periodic assessments would then enable it to pass through a class quicker in spite of its age.

(ii) Mere maturity does not necessarily place a stand in class V. Unthrifty or damaged stands or young stands where the species has failed may all be placed in class V or even in VI if their early felling and regeneration in the next period is desirable.

(iii) If, as is usually reasonable, the final yield is based on the percentage areas in classes IV, V and VI, the critical factor is entry into class IV. Even that point is not highly critical provided that assessments are frequently repeated, say every ten years, or at most 20 per cent of short rotations.

(iv) Even if applied entirely subjectively—in the absence of yield tables, records of age, mixed stands and so on—the forest will be brought steadily towards sustained yields and emphasis can be placed on the urgency of removal and regeneration of individual stands. The method can merely express a restraint, perhaps within a maximum and minimum range, on the quantity of stands to be felled in a period.

B. Volume or volume and increment methods

Although the yield of a forest can be based on its volume and rotation alone, to do so implies use of the increment also. It is better, therefore, to consider all volume methods of yield regulation together.

The regulation of yields by volume of timber cut in order to lead a forest towards sustained yields and normality has long been practised. The origin of the practice can be traced (Dwight, 1965 and Judeich, 1869) to the royal decree of July 14, 1788 issued by the Austrian Government as a Governor's order of August 7, 1788. The decree contained a 'Normale' or standard instruction for the valuation of forest properties and became known as the Austrian Cameral Valuation method.

The *normale* referred only to the equitable valuation of forested lands and *not* to methods of working them. The method adopted in the *normale* was to capitalize the annual money yields that a properly managed forest should be able to provide, using a suitable rate of interest. For the calculation of income it was argued that a properly managed forest, in order to give equal annual yields for the rotation most suitable to the site and conditions, should have a regular series of age-classes which would together contain a definite standard volume of wood. Such a forest with its standard volume of wood would be given the value of the capitalized annual income that it produced. A forest which had been over-cut and therefore contained less than its standard volume would be valued at a proportionally lower figure, and an under-cut forest at a proportionally higher figure. To estimate the volumes of the series of age-classes the volume of a fully stocked mature stand of rotation age was measured and its MAI (volume divided by age) calculated. This MAI was then applied to calculate the volume of all younger age-classes. No allowance was made for thinning

yields so that the annual yield of the standard or normal forest
was the volume of the oldest age-gradation which was taken to
equal the yearly increment of the forest.

(a) The Austrian Formula

By providing the idea and form of a standard or normal forest
at which to aim, the Cameral Valuation method stimulated
foresters to devise methods of so regulating the yield that a forest
would be shaped into its normal state. There arose in consequence
the Austrian formula for yield regulation, a formula based on
the simple and logical principle that if the growing stock of a
forest has a volume greater than its normal volume then it
should be reduced by cutting more than its increment for a time,
and if its volume is too small less than its increment should be
cut to allow the volume to expand.

The Austrian formula was initiated (Dwight, 1965) in a set
of four articles published by C. C. André in 1811 and 1812 in
in the *Economic News*. In 1823 his son, Emil André, published a
book in which he originated the well-known Austrian formula:

$$AY = Ia + \frac{AG - NG}{P}$$

in which AY is the annual yield of a forest, Ia its volume increment, AG is the actual volume of the growing stock, NG is the
volume of the normal growing stock and P is the adjustment
period which Emil André put at rotation length.

The formula evidently incorporates the principle given above
but its successful application depends on how its components
are calculated and also on the state of the growing stock itself.
For example a forest might be composed of a single age-class
of about half rotation age so that its volume equals the normal
volume; the formula then gives the absurd result of a yield equal
to the whole increment although not one mature tree exists.

As pointed out by Dwight (1965) the quantities used in the
formula have undergone almost continuous change since the
formula was first presented. Thus in 1811 a variable period of
adjustment was proposed but Emil André in 1823 proposed
that the rotation should be the period and in 1841 Heyer in his
'modification' returned to the variable period whose length
would be decided according to the circumstances in each forest,

It is not worth while pursuing the history of variations used beyond pointing out the implications involved in different estimates of increment, actual and normal growing stocks and in different adjustment periods.

Increment. The increment used can be estimated in several ways. Originally the MAI of fully stocked rotation-aged stands was used to calculate all the volume terms of the formula so that the increment of the forest would be merely its area multiplied by the MAI per acre of the rotation-aged stands (their standing volume divided by age) and it would also equal the normal increment as then calculated.

That procedure had the advantage of simplicity and continuity of a definite quantity but also the disadvantage that it might differ radically from the true actual increment if many stands were severely depleted and open. So in 1823 André proposed that the actual increment should be used and that has been general practice ever since.

But an important point arises from the method of estimating the actual increment. The earlier method was to calculate the standing MAI of each age-class by felling samples of each age-class and dividing that volume by its age in the same way that the MAI of rotation-age stands was calculated. That method excluded all thinnings from the calculation. *The yield so calculated will, therefore, apply to the final yield only* and the formula so used can be applied only to even-aged forests in which final and intermediate yields can be separated. If, however, the CAI of each age-class is measured (e.g. from borings and measurements of ring width), thinnings are then included and the formula gives the total, global yield in one figure. The formula can then be used for uneven-aged, selection forests also in which final and intermediate yields cannot be distinguished.

Another point may also be noted. A forest having an excessively small growing stock, i.e. with an excess of young stands, will have a high CAI, and a forest with an excess of old stands (which may also be open) will have a comparatively low CAI. Consequently the increment will be higher when the expression (AG—NG) is negative, and will be lower when that expression is positive and high. The CAI will therefore oppose and so delay the rate of adjustment. Additionally the quantity of Ia will vary from period to period. To reduce those two effects von Fiestmantel in 1856 and Gerhardt in 1900 advocated that the increment used should be the mean of the actual (Ia) and normal increment (In). Gerhardt's modification of the Austrian formula (used widely in Germany today) reads:

$$AY = \frac{(Ia+In)}{2} + \frac{AG-NG}{P}$$

The volume of the actual growing stock. Originally the volume of the actual growing stock was calculated by applying the MAI per acre of fully stocked, rotation-age stands to all other age-classes so that their volume equalled their area × MAI × age, irrespective of their stocking. The figure so obtained would be wrong not only because of variety of stocking but also because the process implied that annual increment is uniform to rotation age, whereas in fact it is slow at first, then rapid and then slows down. As the normal growing stock was calculated in the same way, however, there was the great advantage that like was compared with like.

If, on the other hand, allowance is made for poor stocking the value of AG (and consequently also the expression (AG−NG) and the yield, AY) is reduced accordingly so that felling and replacement of old, open stands is delayed in proportion. So the old method had advantages since it accelerated the felling and removal of old open stands although by doing so there would be a tendency also to reduce the average volume (and therefore age and rotation) below normality since some open stocking is inevitable so that AG would always be given too high a figure.

Nevertheless it is reasonable to measure the true volume of AG and then adjust NG as a proper equivalent to AG as suggested below.

The volume of the normal growing stock. As stated earlier the volume of the normal growing stock, NG, was calculated by applying the MAI of fully stocked rotation-age stands to a normal series of age-classes and was comparable with the similarly calculated AG. Later NG was calculated from yield tables. If that is done with no modification NG will be given a volume never attainable in practice since no growing stock can, everywhere, be ideally stocked or ideally healthy. Consequently the expression AG—NG and therefore also AY will always be less than they should be and a forest older and more voluminous than it should be will be attained.

It is therefore necessary to reduce yield tables figures to a stocking comparable with that being attained in practice, perhaps 90 per cent of yield table figures with special allowance for old, depleted stands. Then the expression AG—NG will express correctly the real excess or deficit of AG. Although such a correction of yield table values appears an obvious precaution it has been stressed by only a few authors, e.g. Chapman (1950) although in 1878 the Austrian Government had approved the reduction of yield table figures to average stocking (Dwight, 1965).

To emphasize the importance of applying modified yield table figures in calculating the normal growing stock, the latter might be termed desired growing stock (DG) as proposed by Chapman. The Austrian formula would then read:

$$AY = Ia + \frac{AG-DG}{P}$$

The adjustment period. The longer the adjustment period the smaller is the annual influence of the expression $(AG-NG)/P$ on the annual yield which then depends more on the actual increment, Ia. If there is an excess of old woods Ia will tend to be low and high if there is an excess of young woods—just when the yield should be higher and lower respectively. In such circumstances the slow correction of a long adjustment period will be accentuated by the state of the increment.

A very short adjustment period tends to perpetuate the mal-distribution of age-classes. Thus quick removal and regeneration of an excess of mature and overmature stock causes an excess of young stock; variation of increment with average age of the growing stock then applies a useful correction.

A long adjustment period (e.g. one half or third of the rota-tion) without intermediate revision of the yield allows yields to remain stable during the period but may also allow undesirable changes in the growing stock to occur. Thus an initial excess of middle-aged woods may reach maturity and turn a previous deficit of mature woods into an excess.

One may conclude that the nearer a forest is to normality the longer can the adjustment period be, with reasonably frequent revisions to cope with accidents. In abnormal forests the urgency to remove overmature stands may require comparatively short periods. In broad terms periods of about one quarter to one third of a rotation with revisions of yield every ten to fifteen years are usually satisfactory.

Concluding comments on the Austrian formula. If the formula is judiciously used with frequent revisions of the yield it can give very good results. By using the CAI of the forest, the true volume of the growing stock, a modified volume of the normal growing stock and a reasonably short period of adjustment, a total, global yield is obtained which the forester can apply as conditions require. He can allot the prescribed yield to either thinnings or final yield in the proportions he wishes and select for final felling those stands most urgently requiring it. There is, therefore, great flexibility. But if the forest is very abnormal the use of the formula is best postponed until an area method brings it closer to sustained yield working.

It should not be forgotten that when the CAI is used the formula also gives the total yield and is applicable to selection forests.

(b) Hundeshagen's formula

Hundeshagen assumed that the actual annual yield should bear the same proportion to the volume of the actual growing stock as the normal yield bears to the normal growing stock, so that:

$$\frac{AY}{AG} = \frac{NY}{NG}$$

$$\text{and} \quad AY = AG \times \frac{NY}{NG}$$

The ratio NY/NG expressed as a percentage, $(NY \times 100)/NG$, may be termed the utilization or exploitation per cent. Both NY and NG are taken from yield tables so that the exploitation per cent for a particular forest will be constant, for the species and rotation selected, e.g. 3 per cent. Therefore, to calculate the yield AY all that is necessary is to measure the actual growing stock. There is no need to measure its increment—a great advantage.

The following points may be noted:

(i) As the yield is based on the standing volume of the actual growing stock thinnings are excluded; only the final yield is calculated. (As shown in chapter IV the standing volume of a forest does not include the thinnings already made; to measure the total MAI the volume of thinnings previously made must be added to the standing volume.) Consequently NY must also be the final yield only and NG must be the normal standing volumes of each age-class.

(ii) As the yield is fixed by a percentage of AG the final felling is proportionately greater if there is an excess of AG and *vice versa*. So the general principle of yield regulation is observed. But there is no time limit for adjustment which will be slow. Frequent remeasurements of the AG and revision of AY are necessary.

(iii) As in other formula methods absurd results in abnormal forests are possible. Thus a positive value of AY is always produced whether there is a mature tree in the forest or not.

(iv) The basic assumption is not correct. Young woods grow fast and old ones more slowly so that the ratio AY/AG is not constant in an abnormal forest and therefore cannot equal NY/NG which is constant.

(c) Von Mantel's formula

Hundeshagen's formula can be developed to a simpler state which eliminates the need to consider or calculate a normal growing stock and its increment or yield. Thus by accepting Hundeshagen's thesis we have:

$$AY = \frac{AG \times NY}{NG}$$

$$= \frac{AG \times NI}{NG} \quad \text{because normal increment}$$
(NI) equals normal yield.

But $NI = \dfrac{2NG}{R}$ (See chapter IV)

∴ $AY = \dfrac{AG \times 2NG}{NG \times R}$ (by substitution of terms)

$$= \frac{2AG}{R} \text{ or } \frac{AG}{\frac{1}{2}R} \quad (Von\ Mantel's\ formula)$$

If the exploitation per cent, P, is calculated we have:

$$P = \frac{AY}{AG} \times 100$$

$$= \frac{2AG}{R} \times \frac{100}{AG} \quad \left(\text{because } AY = \frac{2AG}{R}\right)$$

$$= \frac{200}{R} \qquad \text{(often known as } Masson's\ ratio\text{)}$$

To decide the yield, therefore, it is only necessary to measure the growing stock and divide its volume by half the rotation used. The rate of utilization or exploitation is constant for any one rotation, e.g. 2 per cent for a rotation of 100, irrespective of species, site or state of the growing stock.

Implications of the formula. Let us consider the implications of that very convenient state of affairs.

(i) As explained in chapter IV the relation $NI = 2NG/R$ gives the normal final yield only, i.e. the standing volume of the oldest age gradation. Therefore von Mantel's formula excludes thinnings from the yield and is not applicable to irregular, uneven-aged forests. Nevertheless, it has frequently been used for irregular forests that are extensively managed—e.g. in India and Burma, where the yield has been confined mainly to comparatively large trees above a size limit which corresponds with an exploitable age. In those circumstances the yield will be conservative.

(ii) The expression $NI = 2NG/R$ assumes that increment is uniform to rotation age, i.e. that the hypotenuse of the normal forest triangle of volume (see chapter IV) is a straight line, whereas in fact it is sigmoid, growth being slow at first, then fast, then slower and slower.

Therefore Flury's constant, C, should be incorporated in the formula which then reads:

$$AY = \frac{AG}{C \times R} \quad \text{and } not \quad \frac{AG}{\frac{1}{2}R}$$

and
$$P = \frac{100}{R} \times C \quad \text{and } not \quad \frac{200}{R}$$

The constant will vary with species, site and rotation. To calculate values for C the real ratios of NY/NG must be obtained from yield tables.

(iii) The formula will give absurd results in very abnormal forests.

(iv) By using a constant exploitation per cent the formula observes the guiding principle that more should be cut when the volume of AG is in excess and *vice versa*. But there is no period of adjustment which will be slow.

(v) Owing to its simplicity and the need only for a knowledge of the actual growing stock volume it is a very useful formula where yield tables are unavailable and increment is not easily or quickly determined (e.g. for trees which have no annual rings). Also it is useful as a quick and ready check on the yield ascertained by other methods.

Applied examples of the three formulae

The application of the three formulae can be seen from the examples worked out for a hypothetical forest of Douglas fir whose volumes and increment are given in Table 15. The figures used have been taken from the British Yield Tables (1966) for Douglas fir, yield class 200 and applied to an area of 240 acres worked under a rotation of sixty years. The figures given to a normal forest should be as comparable as possible with those of the actual forest, i.e. the normal forest should be based on those densities of stocking that are actually being obtained in practice. So yield table figures per acre have been reduced as explained in the notes below the table. As increment will not fall in proportion to falls in density of stocking a smaller reduction has been used for increment than for volume. The per acre figures shown under 'Standard Values' are then applied to the acreages of *both* the Normal and Actual Forests.

Yield calculations

Using the figures tabulated above the annual yields calculated by the three formulae will be:

(i) *The Austrian formula.* By using the actual CAI of each age-class the annual yield will be global or total and include thinnings and final fellings.

$$AY = Ia + \frac{AG - NG}{P}$$

Table 15

INVENTORY RESULTS FOR A HYPOTHETICAL FELLING SERIES

Douglas Fir—Yield Class 200

Area—240 acres   Rotation—60 years

Age Class (Years)	Standard Standing Volume per acre Hop. ft.	Values CAI. per acre hop. ft.	Normal Forest Standing			Forest Standing		
			Area acres	Volume hop. ft. (NG)	CAI. hop. ft.	Actual Area acres	Volume hop. ft. (AG)	CAI. hop. ft. (Ia)
1	2	3	4	5	6	7	8	9
5	—	—	30	—	—	40	—	—
10	—	—	20	—	—	25	—	—
15	840	187	20	16,800	3,740	20	16,800	3,740
20	1,445	235	20	28,900	4,700	15	21,675	3,525
25	2,015	268	20	40,300	5,360	10	20,150	2,680
30	2,685	271	20	53,700	5,420	15	40,275	4,065
35	3,305	257	20	66,100	5,140	10	33,050	2,570
40	3,655	240	20	72,100	4,800	10	36,550	1,400
45	4,090	221	20	81,800	4,420	5	20,450	1,105
50	4,485	205	20	89,700	4,100	5	22,425	1,025
55	4,870	188	20	97,400	3,760	20	97,400	3,760
60	5,435	172	10	54,350	1,720	25	135,875	4,300
65	5,585	147	—	—	—	20	111,700	2,940
70	5,710	131	—	—	—	10	57,100	1,310
75	5,935	116	—	—	—	10	59,350	1,160
TOTAL			240	601,150	43,160	240	672,800	34,580

Notes. 1. The age-classes are assumed to be the average of stands with ages up to 2½ years on either side of the class age. So the years 1–2½ are not accounted for and class 60 goes 2½ years beyond rotation. So the *normal* area for the 60-year class is 10 acres and for the 5-year class is 30 acres and not 20 acres.

2. Measurement of the growing stock at any time will include *half* the thinning in the standing volume (e.g. if all thinnings are made on a one-year cycle and measurement is made in the middle of the season). So half the thinning volume is included in the standing volume—except in the oldest class of the normal forest and in classes 70 and 75 of the actual forest when the whole thinning volume is included, since thinnings are unlikely to be made in mature stands.

3. The standard values of volume per acre (applied to both normal and actual acreages) have been obtained by reducing yield table values as below:

age-classes   5–35   by 10%
age-classes 40–65   by 15%
age-classes 70–75   by 20% which includes salvage thinnings to remove wind-falls, diseased trees, etc.

4. The normal annual final yield (NY) is taken to be the 60-year age-class volume i.e. the volume on 4 acres each year which equals 5435 × 4 = 21,740 hoppus feet.

5. The increment, i.e. CAI of each class, has been reduced from yield table figures as below:

age-classes   5–60   by 5%
age-classes 65–75   by 10%

Then if the period of adjustment is taken to be 20 years:

$$AY = 34{,}580 + \frac{672{,}800 - 601{,}150}{20}$$

$$= 34{,}580 + 3{,}582$$

$$= 38{,}162 \text{ hoppus feet (total or global yield)}$$

(ii) *Hundeshagen's formula.* Here the annual yield is the final yield only (but see note below the example of von Mantel's formula).

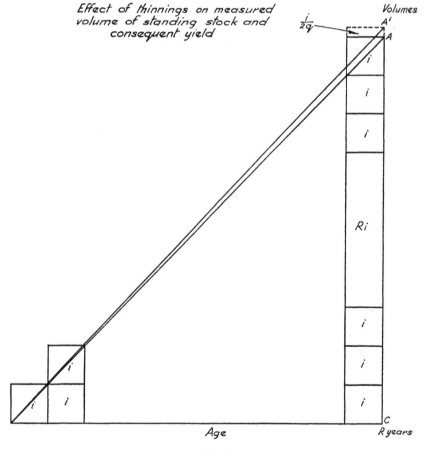

FIG. 7

The exploitation per cent $= \dfrac{NY \times 100}{NG}$ (For value of NY see note (4) above)

$$= \dfrac{21,740 \times 100}{601,150}$$

$$= 3\cdot6 \text{ per cent}$$

$$AY = 3\cdot6 \text{ per cent of } AG$$

$$= \dfrac{3\cdot6 \times 672,800}{100}$$

$$= 24,221 \text{ hoppus feet } (\textit{final yield only}).$$

(iii) *Von Mantel's formula.* Here the annual yield is, very nearly, the final yield only (see note below), namely:

$$AY = \dfrac{AG}{\tfrac{1}{2}R}$$

$$= \dfrac{672,800}{30}$$

$$= 22,427 \text{ hoppus feet (final (approx.) yield).}[1]$$

Von Mantel's exploitation per cent $= \dfrac{200}{R} = \dfrac{200}{60} = 3\cdot33$ per cent.

---

[1] It is not quite correct to say that von Mantel's or Hundeshagen's formulae apply only to the final yield. On average half of the volume being thinned in the current thinning cycle will be included in the measured standing of the growing stock so that the yield will be greater than the true final volume in proportion to that extra volume. But the increase is very small as shown below.

Consider again a normal forest of R age-gradations each 1 acre in extent. Each gradation accrues a uniform increment, i, during a rotation of R years (see Figure 7).

Let us assume that:
 (i) Each of all the gradations is thinned yearly.
 (ii) The amount thinned is $1/q \times i$.
 (iii) The forest is measured in the middle of the growing season in the middle of the thinning programme so that half the stands measured have been thinned and half not. On average, therefore, the volume of each gradation as measured (and corresponding to the main crop after thinning as shown in yield tables) = (its age $\times$ i)$+$i/2q.
the volume of the oldest age-gradation

$$= Ri + \dfrac{i}{2q}$$

Comments

(i) The explanation of the difference in exploitation percentages seen in von Mantel's and Hundeshagen's formulae is that the latter is calculated from yield table figures which give a variable rate of increment. In other words they incorporate Flury's constant which is not incorporated in von Mantel's 200/R which assumes a constant increment. Hundeshagen's formula is, therefore, preferable if yield tables from which to calculate the ratio NY/NG are available.

(ii) It is interesting to note that if the CAI of the normal forest is used for calculating the exploitation percentage (NI/NG) instead of the final yield (NY) used above and as also used by

---

and the measured volume of the whole forest of R gradations, i.e., of $\triangle$ ABC of figure 7

$$= \frac{R(Ri+i/2q)}{2}$$

Now by von Mantel's formula the *yield* of the forest of R acres

$$= \frac{R(Ri+i/2q)}{2} \times \frac{1}{\frac{1}{2}R}$$

$$= Ri+i/2q$$

But the final yield, i.e. the main crop, in only Ri. So the excess is i/2q (per acre of the one age-gradation).

If that excess is applied to our example of Douglas fir, yield class 200, rotation 60 years area 240 acres one can assume that the increment is 200 hoppus feet per acre per year and that q = 2.

The value of $\dfrac{i}{2q}$ then $= \dfrac{200}{4} = 50$ hoppus feet per acre.

The average area of an age-gradation $= \dfrac{240}{60} = 4$ acres. Therefore the

yield over and above the final yield $= 4 \times 50 = 200$ hoppus feet, or less than 1 per cent of the yield given by von Mantel's formula (22,427 hoppus feet).

It may be said that in fact it is the younger stands that are mainly subjected to thinnings and that their increment is much higher than the MAI so that the effect on measurements and the volume of the forest will be greater than shown above. That is true but is counterbalanced by the fact that only about 3/5 of all the age-classes are thinned at all and those near maturity are very lightly thinned or left unthinned. Consequently it is safe to accept the generality that yields calculated by von Mantel's or Hundeshagen's formula are final yields only.

Hundeshagen, a much greater yield (global because it includes thinnings) is obtained, thus:

$$\text{Exploitation per cent} = \frac{NI}{NG} \times 100$$

$$= \frac{43,160}{601,150} \times 100$$

$$= 7\cdot 18 \text{ per cent}$$

$$\text{The } \textit{global} \text{ annual yield then} = \frac{7\cdot 18 \times 672,800}{100}$$

$$= 48,307 \text{ hoppus feet.}$$

(iii) If Gerhardt's modification of the Austrian formula is used we have:

$$AY \frac{(la+NI)}{2} + \frac{AG-NG}{P}$$

$$\text{Then } AY = \frac{34,580+43,160}{2} + \frac{672,800-601,150}{20}$$

$$= \frac{77,740}{2} + 3,582$$

$$= 41,322 \text{ (global yield)}$$

(iv) The reason why the yield is magnified when the normal increment is incorporated in the calculation in Gerhardt's formula and in Hundeshagen's formula is that the actual increment in the example is lower than the normal owing to the preponderance of old age-classes. Thus the actual CAI is delaying the rate of adjustment as explained previously on page 190. Also by reducing yield table volumes by 10 and 15 per cent and increment figures by only 5 per cent the ratio of increment to volume is raised which is to be expected in under-stocked woods.

(d) The regeneration area method

Towards the end of the nineteenth century it was found that the rigidity of the permanent periodic block method was causing disastrous results in the lower and middle montane forests of the Jura and Vosges in France. Conditions there precluded any rigid approach. Although even-aged (regular forests were deemed desirable for convenience of management and for high quality timber), accidental damage from wind, snow and insects is frequently heavy in these mountains and upsets preconceived programmes of work. Moreover natural regeneration, which is greatly favoured, cannot always be obtained quickly and deep snow and short, late springs make artificial regeneration difficult so that regeneration of a block within a fixed time is often impossible and at best unreliable. Variation in sites also causes variation in growth rates so that a fixed, uniform rotation is undesirable.

Consequently there arose the volume method of 1883 which was systematized by Mélard in 1894 as the regeneration area or *quartier bleu* method (not *affectation bleu*).

The distinctive features of the regeneration area method is that compartments, which need not be contiguous, are allotted to three unequally sized sections or areas (*quartiers*); the allotment is revocable at the end of each working plan period.

As in the single periodic block method three sections or areas are allotted, namely:

The regeneration area (the *quartier bleu* since it is painted blue on the maps) which is composed of compartments in which regeneration is in progress or is desirable owing to the ripeness or deterioration of the stands.

The preparation area (the *quartier jaune*, painted yellow on the maps) which is composed of compartments that it is expected will be allotted to the regeneration area at the next revision of the working plan.

The improvement area (the *quartier blanc*, painted white on the maps) which is composed of compartments containing young and semi-mature stands that need weeding, cleaning and thinning.

As the allotment of compartments to the areas is revocable and alterable at each revision of the working plan, the rotation of any compartment can vary, as in the single periodic block method, according to the rate of progress of the stands in it. But in the regeneration area method, unlike the single block method, no time is fixed for completion of regeneration; theoretically a compartment can remain indefinitely in the regeneration area. The regeneration area method is, therefore, even more flexible than the single periodic block method.

The implications that arise from the flexibility which is needed to cope with possibly heavy accidental wind or other damage (*chablis*), to maintain adequate uniformity of stands and to achieve regeneration expeditiously but without a time limit are the following:

(a) *Working plan or management period.* Although there is no period for regeneration, a definite time for work programmes is necessary, i.e. a time at the end of which prescriptions of the working plan are revised. This working plan or management period must be neither too long, whereby prescriptions would become vague and out-of-date, nor too short whereby results of working would not be clear and costs of revision would be repeated too often. Usually the time between revisions varies between fifteen and twenty-five years, conveniently a small multiple of the thinning cycle, reaching thirty years only for very long rotations.

(b) *Size of the areas or quartiers.* (i) The regeneration area. Although the method contains no time limit for regeneration of any particular compartment, the maintenance of sustained yields requires that the area regenerated in a working plan period should average $(A \times P)/R$, where A is the area of the forest or felling series, R is the average rotation or age of exploitability and P is the working plan period. But, to give both flexibility in choice of compartments for regeneration and freedom for regeneration fellings to proceed in harmony with varying conditions, the actual regeneration area should be much larger than $(A \times P)/R$, say half as large again. Then, if the regeneration area $= 1 \cdot 5 \times (A \times P)/R$ for a rotation of 120 and a working plan period of twenty years, 25 per cent of the forest or series would be allotted to the regeneration area. In practice it varies between 20 and 30 per cent of the area of the series.

(ii) The preparation area. This area need not be greater than about half the regeneration area since at the end of the working plan period it is certain that some compartments will still be only partially regenerated and perforce remain still longer in the regeneration area while in some other compartments regeneration may not have begun.

(iii) The improvement area. This comprises the remaining compartments which may include those in which regeneration is almost but not quite complete.

(c) *The prescribed yield.* As the exact area, and therefore the volume and increment of the trees contained in it, which will be regenerated in the working plan period cannot be predicted exactly and as unpredictable losses from wind and other damage must be anticipated, it is essential to use a global or total volume yield. Only then can the extent of all operations be so regulated

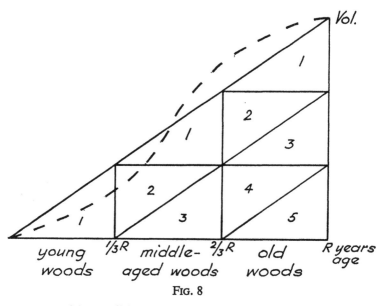

Volumes of a normal even-aged forest cleared by thirds of the rotation

FIG. 8

to cope with conditions as they happen and yet maintain or improve the state of normality of the forest as a whole. Consequently the whole growing stock must be measured and not only that which is found in the more mature compartments.

Although the Austrian formula gives a global yield and can be used for the regeneration area method, the formula usually used is that which arose out of the volume method of 1883.

Consider normal, even-aged forest that has been divided into three equal areas by ages, 1 to $\frac{1}{3}$R, $\frac{1}{3}$ to $\frac{2}{3}$R and over $\frac{2}{3}$R. The normal volume of the growing stock, assuming uniform growth through the age-classes, is then represented by the triangle shown in fig. 8.

The old woods, having a volume OV and annual increment per cent i, would be felled and regenerated in the period $R-\frac{1}{3}R$ years. So the annual yield, by Cotta's formula, would be:

$$\frac{OV}{\frac{1}{3}R}+\frac{1}{2}OV \cdot i$$

The middle-aged and young woods with volumes of MV and YV and an annual increment per cent of i', would be thinned and the amount thinned would be a selected fraction 1/q, of this increment. So the annual yield from them would equal:

$$1/q(MV+YV) \cdot i'$$

The total or global annual yield from all the woods would, therefore, be:

$$\left[\frac{OV}{\frac{1}{3}R}+\frac{1}{2}OV \cdot i\right]+1/q[MV+YV]i'$$

The equilateral triangles in the diagram show that the ratio of the volumes OV, MV and YV are as 5 : 3 : 1, so that OV = 1·67 MV in a normal forest—if growth is uniform. But growth is not uniform so that in fact volumes are more like those included by the dotted line, as has been explained before in chapter IV. The relation OV = 1·67 MV is clearly approximate only and might easily vary from 1·5 MV to 1·9 MV depending on rotation, species and yield class.

Instead of ages let us now consider the diameter of trees and assume that all trees increase uniformly in diameter from youth to exploitable age. We can now classify the growing stock volume according to large trees (LV), medium sized trees (MV) and small trees (SV). The diameter will then respectively be $\frac{2}{3}$ of exploitable diameter and over, between $\frac{1}{3}$ and $\frac{2}{3}$ exploitable diameter and under $\frac{1}{3}$ exploitable diameter.

The small trees (SV) whose comparative volume is small might be excluded from the inventory of the growing stock or at least from the calculation of the yield. Then the global annual volume yield of trees above $\frac{1}{3}$ exploitable diameter would be:

$$\left[\frac{LV}{\frac{1}{3}R}+\frac{1}{2}LV \cdot i\right]+[1/q \cdot MV \cdot i']-\textit{Mélard's formula}$$

That formula, often known as Mélard's formula, is the one commonly used for the regeneration area method.

Comments. (i) The formula really applies only to a normal forest. But it is clear that where there is an excess of old woods and large trees the calculated yield is increased and *vice versa.* But the rate of removing the excess or adjusting the deficit tends to be exaggerated and so may perpetuate the abnormality and be unsuitable in consequence. For that reason and also because of the doubtful assumptions made the yield calculated should be checked by another method such as the Austrian formula and also by comparison of areas occupied by the several age-classes. A compromise total yield may then be adopted which is applied after due consideration of the present state of stands, e.g. the degree of overmaturity and deterioration.

(ii) The two parts of the yield can be applied as the formula suggests—the first portion to felling in the regeneration and preparation area only and the second or thinning portion to the improvement area; wind falls would be debited in the same way according to the place of occurrence. But the usual practice is to cut the global yield in the following order of priorities:

1. Trees damaged by wind, snow, insects, etc.
2. Thinnings.
3. Regeneration fellings in the regeneration area.

If casualties from wind etc. are heavy in any year it is clear that regeneration fellings may get into heavy arrears and even thinnings as well. Some re-adjustment may then be necessary; successive years of heavy casualties might necessitate a re-calculation of the yield or even revision of the plan.

(iii) Recalculation of the yield after remeasurement of the growing stock is advisable after some ten years or at the middle of the working plan period.

(iv) The working plan should indicate the urgency of re-generating individual compartments and provide an order or priority of cutting with an estimate of expected yields during the next five or ten years from regeneration and improvement fellings. Different thinning cycles and different values for $1/q$ can, of course, be adopted for different compartments according to their age and vigour. The sum of the expected yields should be related to the calculated yield and a compromise adopted if necessary after allowance for expected casualties.

(v) The yield prescribed should be a guide rather than an inescapable order on what is to be cut and where it is to be cut, the final decision of which will be decided as current conditions may dictate. The local manager may, however, be confined to quantities within, say, ± 10 or 25 per cent of yields prescribed for any one year beyond which sanction from higher authority would be needed.

(vi) Much depends on the correct measurement of increment, correct choice of the factor $1/q$ and the correct assessment of R which greatly affects the yield and is not easy to determine accurately. Experience and analysis of past yields in relation to the present state of the growing stock will have a great influence on the final decision of what the quantity yield should be and how it should be apportioned.

(vii) It is worth noting that although the method of 1883 divided the rotation into three parts, any other subdivision is possible. If it is divided into four parts the ratio of normal volumes would be $7 : 5 : 3 : 1$ instead of $5 : 3 : 1$.

(viii) As it gives a global yield and is based on the classification of trees by diameter after a full measurement of the growing stock, Mélard's formula can be used for uneven-aged, irregular forests, regular forests and for forests intermediate between the two. That is a considerable advantage in mountainous regions where forests tend to attain no more than a semi-regular structure. But, if the object is to maintain the greatest practical degree of regularity the extreme flexibility inherent in application of the formula must be controlled. The regeneration area method, by allotting compartments to definite treatment regions—of improvement, preparation and regeneration—makes this control possible, while at the same time the regulation of a global yield shapes the volume of the growing stock as a whole by adjustment of excesses or deficits. A fundamental factor is the permanent need to ensure an adequate rate of regeneration, whether natural or artificial.

C. Yield by numbers and sizes of trees

It is obviously possible to regulate the yield of a forest by numbers of trees. If the numbers specified are also designated by size a very detailed control could be provided synonymous with a volume yield allotted not merely to final fellings and thinnings but to all age-classes.

But regulation of yield by numbers of trees is usually used

more extensively, for example in large, irregular tropical forests which are often composed of many species only some of which are marketable and then only those above some fairly large minimum size. Frequently the only practical method of working such forests is to use a fairly long felling cycle, and to confine fellings to trees above a certain size in numbers equivalent to the numbers of smaller, younger trees which will grow to maturity in the next and subsequent cycles. A sustained or increasing yield can then be assured provided that the younger trees are not damaged by the fellings or suppressed and crowded out by unwanted competitors and provided that regeneration continues unchecked. Regulation of the yield is only one factor in assessing or assuring future supplies; silvicultural tending to improve the composition, vigour and form of the younger growing stock remains essential and is expensive if small trees are unsaleable.

As the method adopted by Brandis in 1856 when he organized the management of the Pegu teak forests in Burma displays the principles of regulating yields by number of trees it will be described shortly.

(a) The Brandis method

The method requires a knowledge of three attributes of the growing stock, namely:

> (i) The numbers of trees in each size-class.
> (ii) The times of passage, i.e. the times taken by trees to grow through the various size-classes to exploitable size.
> (iii) The casualty per cent of each size-class, i.e. the per cent of the number of each class that die, are blown over or are thinned before they reach exploitable size.

Having that information it is possible to calculate how many trees reach maturity, i.e. the exploitable size (Class I trees) in any period selected, such as the felling cycle. Then, if the forest is normal, the periodic yield of exploitable sized trees is the number of trees that reach that size during the period. If there is an excess of mature trees then more than that number should be cut temporarily and, conversely, less should be cut if there is a deficit.

Working stock

But there is an additional factor. There must be a sufficient stock of exploitable sized trees on the ground at the start of the period or felling cycle. The reason becomes clear if we consider a simple example:

Consider a normal, uneven-aged felling series in which:

(i) The felling cycle is ten years.

(ii) The time of passage from class II to class I trees is, say, twenty years (but does not concern us at present).

(iii) The annual yield is forty class I trees or 400 during the cycle.

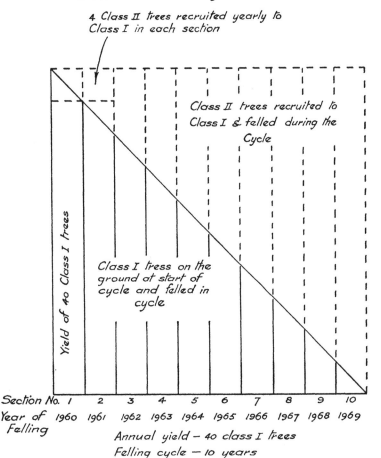

Working stock of Class II trees needed to maintain a constant yield of class I trees

4 Class II trees recruited yearly to Class I in each section

Class II trees recruited to Class I & felled during the Cycle

Class I trees on the ground at start of cycle and felled in cycle

Yield of 40 Class I trees

Section No.   1   2   3   4   5   6   7   8   9   10
Year of Felling   1960 1961 1962 1963 1964 1965 1966 1967 1968 1969

Annual yield – 40 class I trees
Felling cycle – 10 years

FIG. 9

So 400 trees must grow into class I from class II during the cycle, i.e. forty trees each year throughout the whole felling series. Therefore four trees move up yearly in each of the ten sections.

Now in year one of the cycle, in section I, there is no time for trees to grow from class II to I. There must already be enough class I trees on the ground in section I to supply the yield of 40 trees.

So, taking average conditions, i.e. the middle of the growing season when half the annual recruitment of four trees to Class I in section I is finished, there must be thirty-eight class I trees already on the ground to which two trees (half the annual recruitment from class II) can be added in the second half of the growing season to make up the yield of forty trees. And similarly through the other sections year by year so that after ten years in section 10 only two class I trees need to have been on the ground at the start of the cycle since thirty-eight recruits from class II will become available during the cycle. It is clear, therefore, that half the yield during the cycle is composed of trees that were already class I trees at the start of the cycle. This is illustrated in Fig. 9.

So, in general, the stock of class I trees available on the ground as working stock at the start of a cycle must equal half the recruitment from class II during the cycle.

Calculation of the yield

The procedure for calculating the yield is best explained by an example. Table 16 shows the (imaginary) results of an inventory of teak trees in a teak felling series in Burma.

The figures used in the calculation will be the times of passage (col. 4) and the number of trees that will eventually reach exploitable size (col. 6). From them are calculated the average rate of recruitment to class I and the consequent numbers of working stock (class I trees) needed.

But the rate of recruitment to class I will vary from cycle to cycle and at least two rates can be considered here, namely the average rate of recruitment for 122 years (from trees sized $1\frac{1}{2}$ to 7 feet) and the rate during the first cycle. In the example those two rates of recruitment are:

(i) Average over 122 years $= \dfrac{100,644}{122} = 825$ trees annually

(ii) During first cycle of 30 years $= 15{,}589 + \left(\dfrac{3}{31} \times 24{,}815\right) \div 30$

$$= 600 \text{ trees annually}$$

Two calculations of the yield are then available:

(a) Using the average recruitment over 122 years.

Necessary working stock $= \dfrac{825 \times 30}{2} = 12{,}375$ trees.

Excess working stock $= 25{,}412 - 12{,}375 = 13{,}037$ trees.

Then, if the excess stock is removed in two felling cycles, the

Annual yield $= 825 + \dfrac{13{,}037}{60} = 1042$ *class I trees.*

Table 16

TEAK GROWING STOCK IN A FELLING SERIES

Felling cycle = 30 years

Class and girths (feet)	No. of trees	Average age of lower limit of class (yrs.)	Time of passage (years)	Casualty% before final exploitation	No. of trees finally available for exploitation
1	2	2	4	5	6
I 7 & up	26,750	152	27	5	25,412
II 6–7	18,340	125	31	15	15,589
III 4½–6	35,450	94	32	30	24,815
IV 3–4½	63,360	62	32	50	32,680
V 1½–3	110,240	28		75	27,560

TOTAL 122

(b) Using the recruitment during the first cycle.

Necessary working stock $= \dfrac{600 \times 30}{2} = 9000$ trees.

Excess working stock $= 25{,}412 - 9000 = 16{,}412$ trees.

Then, if the excess stock is removed in two cycles, the

Annual yield $= 600 + \dfrac{16{,}412}{60} = 874$ *class I trees.*

Comments

1. The average rate of recruitment over a long period would seldom be used owing to the inevitable errors attached to the estimated casualty percentages of the smaller size classes.

However, it is a useful guide to future trends; but the rate of recruitment in the second cycle is likely to be more useful. In our example the recruitment in the second cycle would be 815 trees compared with 600 in the first cycle annually, thereby indicating that yields will tend to rise so that there is no need to husband overmature stock.

2. The decision whether to remove excess stock of class I trees quickly (in one cycle) or slowly (in two or more cycles) depends on the degree of overmaturity of the trees and also on the need to conserve them to make good a possible shortage of class III trees which would involve a fall in yield in the second cycle. In our example recruitment tends to increase materially in the second cycle so that the yield of calculation (b) might reasonably be increased from 874 to 1000 trees yearly, a figure which is also indicated by calculation (a). (A yield of 1000 trees would imply removal of working stock in about forty years instead of sixty years.)

3. The annual yield can also be expressed as a per cent (P) of the class I trees available on the ground, thus:

$$P = \frac{Y}{I + \dfrac{R}{2}} \times 100,$$

where R = recruitment to Cl. I in the cycle and Y = yield in the cycle.

In our example, calculation (b)

$$P = \frac{874 \times 30}{25,412 + 9000} \times 100 = \underline{\underline{76 \cdot 3 \text{ or say } 75 \text{ per cent.}}}$$

Then the yield for the year being 874 trees, the trees selected for felling are chosen on the ground by marking three out of every four class I trees and leaving one. This procedure distributes the yield uniformly over the area and simplifies silvicultural decisions on what should be left.[1]

4. When a minimum exploitable size is used for limiting

[1] The Smythies' (1933) formula uses this percentage but, as first proposed and as applied in India in recent years, the yield is based only on the number of recruits to class I with only a subjective allowance for excess or deficit working stocks. Provided that the forest has been carefully managed for many years so that the size-classes are not too abnormally distributed, no harm need arise. But if excesses or deficits of stock are not considered there is always a likelihood that yields will fluctuate.

fellings and defining yields, the average size of trees felled (in a normal forest) will be that minimum size *plus* half the average increase in size during the cycle.

5. By confining the yield (and fellings) to one or more selected species (such as teak in the example) and by allowing other less valuable species to remain, conditions for the selected species will deteriorate. It is essential for undesirable stems and species to be removed as well (as thinnings), both those above and below exploitable size. Thinnings in smaller size-classes are not included in the prescribed yield but should be made and their quantity included in the casualty percentages.

6. Heavy damage of smaller class trees is liable to occur when felling large tropical trees (Dawkins, 1958). This damage should also be included in the casualty percentages and may cause a severe waste of stock.

7. Yields should be re-assessed at least at the end of each cycle and preferably not less often than every fifteen or even ten years, e.g. at mid-cycle. Records of trees felled and of inventories made and their comparison with subsequent inventories will lead to more exact determination of casualty percentages, subsequent yield determination and success of silvicultural methods.

(b) The method of control

The method of control, which is properly only applicable to uneven-aged forests worked under the selection system, was first proposed in France by Gurnaud and then developed by Biolley and Favre in Switzerland where it is mainly practised and is particularly suited to the conditions.

Essentially the method is not one of yield regulation but rather a procedure of repeated inventories and continuous records which are entirely comparable so that a complete history of the growing stock, its volumes and distribution of size-classes, yields, casualties and increment, is maintained for each working unit of the forest and the forest as a whole. The progress and output of the forest can then be related to each other and to the treatment applied and used to define future treatment and yields. In essence, therefore, the method of control is a method of measurement, accountancy and record to supply facts on which to plan action.

The basic features of the procedure used in Switzerland are designed to ensure true comparability of the measurements and the estimates of volume. They are summarized below:

(i) Total inventories of the whole forest are made at short intervals, such as five to eight years, equivalent to the felling cycle.

(ii) Trees are measured only at breast height, at 1·3 metres.

(iii) The point of measurement is marked with a scribe-cut to ensure subsequent measurement at the same spot.

(iv) Measurements are classified in 5 centimetres diameter classes, usually by species.

(v) Volumes of trees are estimated by using a standard, local volume tables based on diameters at breast height (DBH) only. The *same* table is used for all subsequent inventories.

(vi) Whenever a tree is felled, is wind-blown or dies it is measured at the scribe mark and its volume calculated in the same way.

(vii) The figures are classified for the forest as a whole and for each working unit which is usually the compartment.

(viii) Increment is calculated by each diameter-class or more usually each main group of diameter-classes for the forest as a whole and each working unit by the formula $I = V_2 + E - V_1$, where V_1 and V_2 are the volumes at the first and second inventories respectively and E is the volume of casualties and trees exploited.

(ix) To distinguish the *estimated* standing volumes from the true volumes measured after felling, the standing unit of volume is called a silve and not a cubic metre. The silve is, therefore, the management unit of volume and the cubic metre the commercial sale unit. The relation between silve and cubic metre can, of course, be determined empirically.

Application of the method of control

The basic information wanted is fourfold:

(i) The numbers and volumes of trees in each diameter class, i.e. the stem frequencies and their volumes.

(ii) The numbers and volumes of trees felled (including casualties) in each diameter class, i.e. the yield.

(iii) The increment of each diameter class or group of classes. This is obtained from items (i) and (ii) from successive inventories and is dependent on the time of passage through each class.

(iv) The desirable stem frequencies (and consequent volumes) of a growing stock that is in a state of equilibrium when providing the maximum desirable yield—i.e. the structure which produces steadily the greatest value in increment (and hence yield), an increment and yield which are constantly renewed by regeneration and, from it, growth through each diameter class to the desired maximum size.

That desirable structure can be ascertained only after prolonged management and study of results shown by the successive inventories. But that is not enough. Weather conditions affect increment and may vary in cycles and thereby disturb the interpretation of results. So a fifth item of information is also needed, namely:

(v) A record of the local weather, particularly temperatures, humidities and atmospheric precipitations.

Calculation of increment and times of passage

An example of the calculation of increment for the three groups of diameter classes usually used in Switzerland by the formula given earlier ($I = V_2 + E - V_1$) is shown in Table 17.

For simplicity only numbers of trees have been shown in the table. The increment of the large tree group is the volume of the trees in column 6 ($V_2 + E$, but without the fifteen trees of 55 centimetres diameter which have risen from the medium tree class) *minus* the volume of trees in column 2 (V_1). Similarly the increment of the medium tree group is the volume of the trees in column 6 (which includes the fifteen trees which have progressed to the large tree group and are now 55 centimetres diameter and excludes the fifty-eight trees recruited from the small tree class). As a check the totals of V_1 in column 2 must equal the totals of $V_2 + E$ in column 6.

The time of passage through each diameter class is calculated from the percentage or proportion of stems that rise from a class to the next higher class during the interval between inventories. Then:

$$\text{Time of passage} = \frac{\text{No. of stems in class}}{\text{No. of stems rising}} \times \text{length of interval}$$

$$\text{The average diameter increment then} = \frac{\text{diameter class interval}}{\text{time of passage}}$$

If the interval between two inventories is so long that stems rise through more than one class, two diameter classes may be grouped into one.

Table 17

INCREMENT CALCULATION BY SIZE GROUPS, ACCORDING TO GURNAUD

(from Hermann Knuchel, Planning and Control of the Managed Forest, translated by Mark L. Anderson (1953), First English Edition, Oliver and Boyd, Edinburgh, by permission of the authors and publishers).

Size (Diam.) Class cm.	Number of stems				The stems of V_1 appear again in V_2+E, increased by their increment	Recruitment to the Upper Group
	V_1	V_2	E	V_2+E		
1	2	3	4	5	6	7
85	1	1	—	1	1	—
75	—	2	1	3	3	—
70	3	5	1	6	6	—
65	16	11	—	11	11	—
60	21	13	2	15	15	—
55	51	27	3	30	15	15
Large trees	51	59	7	66	51	
Recruitment into the large tree group					15	—
50	40	32	3	35	35	—
45	55	66	11	77	77	—
40	82	69	11	80	80	—
35	115	117	26	143	85	58
Medium trees	292	284	51	335	292	
Recruitment into the medium tree group					58	—
30	188	142	40	182	182	—
25	219	199	60	259	259	—
20	215	215	25	240	123	117
Small trees	622	556	125	681	622	
TOTALS	965	899	183	1082	965	
Recruitment into the main stand					117	
					1082	

An example of a calculation of times of passage and diameter increments for each class is shown in Table 18. It may be noted that in this calculation, as in the previous one of volume increment, the diameter classes are recorded and the calculations made in descending order of magnitude.

The table shows that in 1910 there was one 90 centimetres tree (col. 3) which must have risen from the one 90 centimetres tree of 1904 in column 2. So one tree is entered as rising in column 5 in the 90 centimetres row and no stem stationary

Table 18

CALCULATION OF THE TIME OF PASSAGE AND DIAMETER IN-
CREMENT BY 5 CM. DIAMETER CLASSES

Data from H. Biolley's, L'Aménagement, p. 51 (after A. Meyer , H. Knuchel, 1950, Planung und Kontrolle im Forstbetrieb, Verlag Sauerlander, Aarau, by kind permission of the publishers.

Diam. class cm.	1904	Number of stems 1910 plus ex- ploited stems	Stems stationary in the class	Stems rising into higher class Number	%	Time of passage through class years	Annual Diam. incre- ment mm.
1	2	3	4	5	6	7	8
100	—	1	—	—	—	—	—
95	—	—	—	—	—	—	—
90	1	—	—	1	100	6·0	8·3
85	1	3	1	—	—	—	—
80	4	5	2	2	50	12·0	4·2
75	7	11	4	3	43	13·9	3·6
70	12	19	5	7	58	10·4	4·8
65	23	34	9	14	61	9·9	5·1
60	41	56	16	25	61	9·9	5·1
55	64	87	24	40	62	9·7	5·2
50	112	166	49	63	56	10·7	4·7
45	219	265	102	117	53	11·3	4·4
40	308	349	145	163	53	11·3	4·4
35	394	515	190	204	52	11·5	4·3
30	613	685	288	325	53	11·3	4·4
25	772	969	375	397	51	11·8	4·2
20	1163	1256	569	594	51	11·8	4·2

687 = RECRUITS INTO MAIN STAND

(col. 4). For the 85 centimetres row, since no stem has risen to 90 centimetres (col. 3) the entry in column 5 is none and one stem is stationary (col. 4); therefore two stems have risen to 85 centimetres and are recorded in column 5, and so on. Columns 6, 7 and 8 are self-explanatory.

The stem frequencies and volumes

The value of records to show the progress and output of a forest, i.e. the results of treatment, is well shown in the two

Tables 19 and 20 and their illustrative graphs in Figs. 10 and 11. The figures were kindly supplied by L. A. Favre in personal correspondence with the author and apply to the initial area of 55 hectares of felling series No. 1 of the forest of Couvet, Switzerland; additions to the area of forest since 1890 are excluded so that figures are comparable.

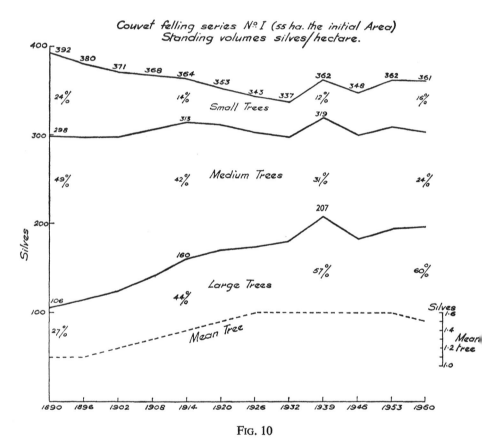

FIG. 10

The figures are for the whole felling series so that in themselves they do not express the actual irregular, uneven-aged character of the forest; one compartment might, for example, be composed entirely of small trees, another of medium trees and so on. It is essential for each compartment to contain the full range of diameter classes. Similar records, therefore, must be maintained for each compartment to guide its treatment so that each

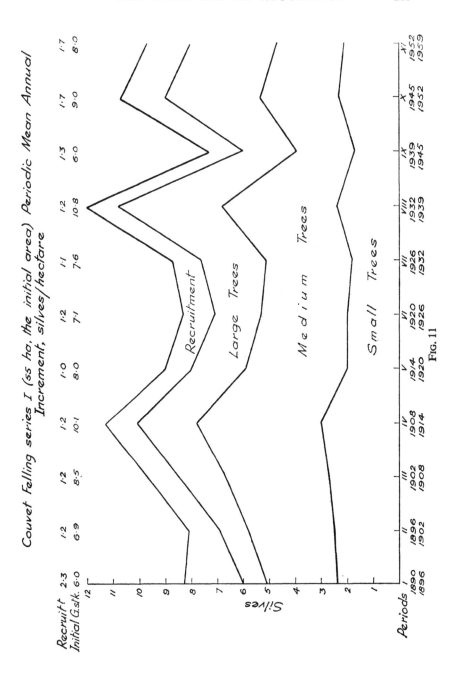

Couvet Felling series I (55 ha, the initial area) Periodic Mean Annual Increment, silves/hectare

Fig. 11

Table 19

FOREST OF COUVET, FELLING SERIES NO. 1 (INITIAL 55 HECTARES). RESULTS OF INVENTORIES

The Standing Growing Stock

(Silves/hectare and percent of total volume in brackets)

(Small trees 20, 25, 30 cm. DBH., Medium trees—35, 40, 45 and 50 cm, Large trees—55 cm. and over)

	Year of Inventory											
	1890	1896	1902	1908	1914	1920	1926	1932	1939	1946	1953	1960
Total volume (and %)	392(100)	380	371	368	364	353	343	337	362	348	362	361(100)
Small trees (and %)	94(24)	83(22)	74(20)	63(17)	51(14)	42(12)	41(12)	40(12)	43(12)	49(14)	54(15)	58(16)
Medium trees (and %)	192(49)	183(48)	174(47)	165(45)	153(42)	141(40)	130(38)	118(35)	112(31)	97(28)	94(26)	87(24)
Large trees (and %)	106(27)	114(30)	123(33)	140(38)	160(44)	170(48)	172(50)	179(53)	207(57)	202(58)	214(59)	216(60)
Volume of mean trees	1·1	1·1	1·2	1·3	1·4	1·5	1·6	1·6	1·6	1·6	1·6	1·5

Table 20

Periodic Mean Annual Increment and Exploitation (Silves/hectare/year)

	Periods										
	I	II	III	IV	V	VI	VII	VIII	IX	X	XI
Small trees	2·4	2·5	2·7	3·0	2·0	2·0	1·8	2·4	1·7	2·3	2·1
Medium trees	2·7	3·3	4·0	4·8	3·9	3·3	3·3	4·4	2·2	3·0	2·6
Large trees	0·9	1·1	1·8	2·3	2·1	1·8	2·5	4·0	2·1	3·7	3·3
Total initial growing stock	6·0	6·9	8·5	10·1	8·0	7·1	7·6	10·8	6·0	9·0	8·0
Recruitment	2·3	1·2	1·2	1·2	1·0	1·2	1·1	1·2	1·3	1·7	1·7
Total increment	8·3	8·1	9·7	11·3	9·0	8·3	8·7	12·0	7·3	10·7	9·7
Increment % in initial growing stock	1·6	1·8	2·3	2·7	2·2	2·0	2·2	3·2	1·7	2·7	2·2
Exploitation	10·5	9·4	10·3	11·7	11·0	10·2	9·8	8·3	9·4	8·5	10·0
Exploitation % on initial growing stock	2·7	2·5	2·8	3·2	3·0	2·9	2·9	2·5	2·6	2·4	2·8

working unit or compartment is also a sustained yield unit and a microcosm of the whole forest or series. That indicates the intensity of management needed to apply the selection system and method of control.

Points of interest in the figures are:

(i) The total volume of the growing stock fell fairly rapidly from 392 silves/hectare in 1890 to 337 silves in 1932 during the initial establishment of true selection forest then rose to 362 in 1939 and, in 1953 and 1960, has been nearly static at 362 and 361 silves/hectare. A volume of about 360 silves/hectare may be near the ideal total volume of growing stock above 20 centimetres diameter. Taking a silve to be 1 cubic metre, the volume is about 4000 hoppus feet or 5000 cubic feet/acre.

(ii) In spite of the fall from the initial total volume, the volume of large trees has risen steadily from 106 to 216 silves/hectare in 1960 and the percentage of large trees from 27 to 60 per cent which may be close to the ideal for the exploitable sizes wanted. The value of the growing stock must have risen greatly in spite of the fall in its total volume since the value of large trees per unit volume must be substantially higher than that of small trees.

(iii) The increase in large trees is balanced mainly by the fall in the volume of medium trees, namely from 192 silves/hectare and 49 per cent of the total volume in 1890 to 87 silves and 24 per cent in 1960.

(iv) The volume of small trees also fell at first, from 94 silves/ hectare and 24 per cent in 1890 to 40 silves and 12 per cent in 1932 since when the figures have risen steadily to 58 silves/ hectare and 16 per cent in 1960. If the figures had continued to fall after 1932 it would have been clear that the growing stock was merely getting older at the expense of regeneration and recruitment to the main stand above 20 centimetres diameter. That this was a danger is shown in Table 20 which shows that recruitment was falling steadily from 2·3 silves/hectare in period I (1890–96) to 1·1 silves in period VII (1926–32). After period VII it has risen steadily to 1·7 silves in periods X and XI (1946–60), during which time the volume of medium trees was considerably reduced from 130 silves/hectare to 87 silves in 1960. It is clear that the correct proportions of diameter classes and the total volume are delicate matters.

(v) The volumes of the mean tree, related to the total volume of the growing stock is a useful indicator of the structure of the growing stock. It may be that a mean tree of 1·5 silves in a growing stock of 360 silves/hectare is about right for Couvet, felling series No. 1.

(vi) Exploitation has been remarkably steady being 10·5 silves/ hectare in period I and 10·0 silves/hectare in period XI, never falling below 8·3 silves/hectare, in period VIII when growing stock was allowed to accumulate by 25 silves/hectare. The exploitation per cent has also been steady and appears stable at about 2·5 or 2·6 per cent.

(vii) In recent years total increment and exploitation have been rougly related. When total increment has risen materially (e.g. to 12·0 silves/hectare in period VIII) the exploitation has been increased in the next period (e.g. from 8·3 to 9·4 silves/ hectare in period IX). Conversely when increment has fallen (e.g. to 7·3 silves/hectare in period IX) exploitation has been reduced in the next period (e.g. to 8·5 silves/hectare in period X).

(viii) The somewhat pronounced variation of increment in different periods, clearly illustrated by the peaks and depressions of fig. 11, cannot be attributed only to treatment which has followed a steady policy. The variations which have an obviously disturbing effect on management must depend largely on cyclic weather conditions.

L.-A. Favre has explained in correspondence with the author that the period IX (1943–46) suffered from an unfortunate and unusual combination of meteorological events which must have severely debased timber increment. At first there were two or three wet years but their beneficial effect was nullified by low temperatures. Then there were three or four years which were not only dry but hot. Only one year was damp and warm. In particular during the months of April to June, the time when woody increment is accumulated, the lack of moisture was particularly striking in 1943, 1944 and 1945 while temperatures were extremely high from 1942 to 1945.

The figures in Table 21, supplied by L.-A. Favre, show the variation of mean annual temperature and precipitations from the means of the last hundred years.

Intensive management under the method of control clearly requires accurate records of weather conditions and more detailed research on the effect of weather on increment is urgently needed. To this end at Couvet in 1953 two forest meteorological stations were set up in places where inventories which classify trees in 5 millimetres instead of 5 centimetres diameter classes are being made for an intensive study.

Table 21

TEMPERATURES AND PRECIPITATIONS AT COUVET

Year	Precipitation change		Temperature change	
1939	Very wet	+500 mm.	Cold	−0·3°C
1940	Wet	+200 mm.	Cold	−0·7°C
1941	Dry	−120 mm.	Cold	−0·5°C
1942	Dry	−120 mm.	Normal	−0·1°C
1943	Dry	−100 mm.	Hot	+1·3°C
1944	Wet	+120 mm.	Hot	+0·3°C
1945	Dry	−200 mm.	Hot	+1·0°C

(Figures rounded off to about 20 mm.)

Yield regulation

The method of control obviously gives full opportunity for delicate adjustments of the annual yield to lead the forest towards the state which best satisfies the objects of management. The accumulated records make it possible to estimate fairly closely what the ideal or normal growing stock volume should be. As yield tables for uneven-aged forests are not available (except as records of stands of individual compartments considered normal) such an estimate would otherwise only be possible by applying figures from even-aged yield tables on the unjustified assumption that the same volumes and the same proportions of age-classes (and their equivalent size) occur in selection forest.

But the method of control provides, in due course after some years, a value for the normal growing stock volume and also a knowledge of its actual volume and increment. So the yield can be calculated by the Austrian formula (see p. 189):

$$AY = I + \frac{AG - NG}{P}$$

The calculation should be applied separately to each compartment since each compartment ought to be completely irregular and in dynamic equilibrium as a sustained yield unit.

But the yield calculation should be used only as a guide. The actual amounts cut should be decided on the condition of

individual trees, the local density, proportions and inter-
mixture of diameter classes and the silvicultural appreciation
of the entirety and the state of the regeneration in each locality
or stand in a compartment at the time of marking. The stem-
frequency curve for each compartment and its variation from the
curve deemed to be ideal will strongly influence the incidence of
cutting in the various diameter classes and the consequent
volume yield.

When a yield is prescribed at the revision of a working
plan the best procedure is for the working plan officer to estimate
the yield, preferably by diameter classes or groups of diameter
classes, of each compartment according to his own assessment.
The subsequent volume can then be expressed as an exploitation
per cent of the standing growing stock volume—e.g. 2·5 per cent
annually or 15 per cent for a six-year period. That yield will
then be both a guide to the officer marking the actual cut later
when the compartment is worked according to the sequence
prescribed in the felling plan, and also a forecast of the probable
outturn of the forest for commercial purposes.

General comments on the method of control

(i) The practice of total inventories at comparatively short
intervals of time introduces an extra cost and complication of
organization. The intensity of management and its cost can be
justified only if results show comparable benefits particularly
in increased values of increment and yield which would not be
attained or maintained by using less intensive methods. For an
irregular silvicultural system, particularly the true selection
system which it is difficult to apply and assess without detailed
figures and records for small units of working, the method of
control is probably the only satisfactory method. In hilly country
with a rigorous climate an irregular system, preferably the true
selection system, may be essential for protection of the soil and
security of the growing stock against wind damage. There is
then little argument against the selection system and the method
of control although high intensity of working and costs may
make it difficult to apply them in large forests where labour
is insufficient or expensive.

That disadvantage makes the use of a sampling procedure
instead of total enumerations attractive. But to avoid sampling
errors which, even if only ±1 per cent will gravely disturb

assessments of increment which may be only 2 per cent of standing volumes, only large permanent sample plots representing 15 to 20 per cent of each working unit are likely to be acceptable. Provided that the actual treatment inside and outside permanent plots in each compartment remains the same, such permanent samples might prove a desirable and satisfactory substitute for total inventories.

(ii) The method of control uses volume estimates based only on stem diameter measurements and a local volume table. There is no consideration of height or form factor the accurate measurement of which is less easy and would introduce an additional and variable source of error. For selection forests in which the average age of separate units does not change once equilibrium has been established and in which site quality is not too variable, that procedure is satisfactory. But for more even-aged forests the average age of individual stands and compartments will change materially and their form factor may also change and differ variously from that of the average form factor incorporated in the volume table for the various diameter classes. Consequent calculations of increment may then not be wholly comparable particularly in a large forest with many site qualities in which the form factor may change differently with increasing age of tree. If that is true then the use of the method of control for a *femelschlag* system of silviculture may be inadvisable and even more inadvisable for true even-aged systems.

(iii) Nevertheless repeated inventories to record the history of the standing growing stock, in volume, increment, age—or size-class structure, and of its yields are essential for the reasoned management of any forest. Only then can a true unbiased assessment be made of its growth and yield capacity, the success of treatment, the performance of various species in various sites or its commercial success. Experimental research for selected problems is valuable, but an intensively managed forest requires continuous, reliable records of treatment accorded and consequent growth and yields from each working unit. Eventually working units should coincide with at least the main site types in each of which will be grown the most suitable forest types each with its sequence of stand types according to age and treatment. If random sampling procedures particularly with a low intensity are used, successive comparisons

8

of increment in each unit are unlikely to be reliable. But the changes in growing stock volume and increment over a prolonged period will be reliable while records of yields, probably based on sales, are likely to be very reliable. A 'historical' increment and exploitation per cent can in time be established separately, if possible, for each main site type.

II. SUSTAINED YIELDS NOT PARAMOUNT

The methods of yield regulation described hitherto have all been based on the assumption that sustained yields and their associated normal growing stocks have been an essential object of management. But a sustained yield may not be wanted or may be subsidiary to one or more other objects. A different approach to yield regulation is then needed, but the approach will still be one that strives to lead the forest to a normal or ideal state; but that ideal state will differ from the classic normal forest which provides sustained yields.

It is therefore essential to define what the ideal state of the forest is and to do that it is first necessary to decide what the purpose of management is and the relative importance of several objects of management. To simplify and clarify that decision it is advisable to divide the forest into working circles in which the purposes of management are distinct. Thus separation, in however dispersed a manner, of areas suitable for recreation and aesthetic enjoyment from areas that can be devoted to commercial forestry enables two distinct working circles to be formed so that treatment of each is freed from constraints imposed by the other; each has a simpler and clearer purpose, each a simpler ideal state.

Although the separation of working circles simplifies the objects pursued in each and facilitates their management, it is certain that some constraints on the pursuit of one main object will arise to some degree. Thus a yield sustained to some extent may be desirable even if the prime purpose is maximum profit. Markets too may seriously affect quantities and species felled and the current performance of individual stands will always affect decisions on what to cut. Yield regulation will, therefore, depend on a clear definition of the purpose of management for each working circle and a consequent choice of those stands which ought to be felled finally and regenerated (and those to be thinned or otherwise treated and the intensity) in the next period (say ten years) to achieve that purpose in the conditions obtaining at the time.

The stands chosen for final felling should be classed in the order of priority or urgency. Then the actual amount to be felled and regenerated can be adjusted to conform with constraints imposed by subsidiary objects or other factors such as markets or funds and labour available.

In effect that would involve allotting to each felling series a regeneration area, a thinning area and an establishment area (weeding and cleaning) similar to the allotments in the single periodic block and regeneration area methods. But the *rate* of regeneration would not be influenced primarily by the distribution of age-classes but on the urgency of final fellings to satisfy the prime object of management within the limits imposed by the constraints.

Hypothetical example

Let us consider a simple hypothetical example with the data shown below:

Area of felling series—2520 acres

Objects of management. (i) Prime object—attainment of the maximum profit, defined by maximum NDR per £100 invested. (ii) Secondary object—provision of regular local employment which would require reasonably sustained yields.

Silvicultural system—even-aged coniferous high forest with mainly clear-felling and artificial regeneration; conversion of existing oak high forest to conifer. Three main coniferous species with an expected average rotation of about sixty years.

Organization—Working plan period—ten years. Growing stock classified by stands in sub-compartments, and grouped in three, dispersed sections—a regeneration area, thinning area and establishment area. The regeneration area is composed of those stands whose felling and regeneration (with the same or another species) during the next ten years is desirable to increase their NDR (see Appendix III(*c*) for an example and chapter IV, p. 71) arranged in the order of priority of felling. The thinning area will be thinned at the best economic intensity to give a sustained output without loss of increment. Areas allotted:

Regeneration area	530 acres
Thinning area	1430 acres
Establishment area	560 acres
	2520 acres

Table 22 summarizes the areas in the regeneration area by priority for felling and shows that the *desirable* yield over the next ten years (to satisfy the prime object of management) = 2,090,000 hoppus feet obtained by felling and regenerating an area of 530 acres, namely 195 acres of oak and 335 acres of conifer.

Table 22

SUMMARY OF GROWING STOCK IN THE REGENERATION AREA

Priority for felling	Location by sub-compts.	Area (acres)			Volumes (h. ft.) available for felling in period (standing vol.$+\frac{1}{2}$ increments).		
		oak	coni-fer	total	Oak	Conifers	Total
1. Very urgent	31 (*a*), (*c*) etc.	55	20	75	120,000	95,000	125,000
2. Urgent	42 (*d*), 56 etc.	75	75	150	195,000	325,000	520,000
3. Very desir-able	28 (*b*), (*c*) etc.	35	145	180	110,000	715,000	825,000
4. Desirable	15 (*d*+*c*), 63 etc.	30	95	125	80,000	450,000	530,000
Totals		195	335	530	505,000	1,585,000	2,090,000

Constraints

It is now necessary to consider the constraints which may require the desirable yield to be altered. There are three essential constraints that affect the final yield:

(i) *Subsidiary object.* To improve age-class proportions it is decided that the final yield over the next ten years should be within +50 or −20 acres of the 420 acre 'normal' cut (for a rotation of 60 years), i.e. the ten-year area yield should be within the range of 400 to 470 acres.

(ii) *Market limits.* Conifer and oak are affected differently, as below:

(a) Oak. There is no commitment to supply oak and the market is weak and sensitive to overloading. Therefore it is decided to limit the felling of oak during the next ten years to 400,000 hoppus feet or about 160 acres in approximately uniform yearly amounts.

(b) Conifers. The market is firm and likely to remain so. No maximum limit to fellings is therefore necessary. But agreements have been concluded to supply 1,100,000 hoppus feet in the next ten years to a local sawmill, equivalent to clear felling 230 acres, in approximately equal yearly amounts.

Therefore constraints imposed by markets are:

(1) Oak. Maximum area yield = 160 acres in ten years.
(2) Conifer. Minimum area yield = 230 acres in ten years.

(iii) *Limits of finance and labour.* It is decided that resources cannot cope with the regeneration of more than 450 acres in ten years in roughly equal yearly amounts.

Final yield calculation

Constraint no. 3 limits the ten-year yield to 450 acres. That is within constraint no. 1 (yields between 400 and 470 acres) and less than the desirable yield of 530 acres.

Therefore the total area yield is fixed at 450 acres in ten years.

It remains necessary to allot the 450 acres to oak and conifer stands.

As the area of oak felled must not exceed 160 acres (constraint no. (ii) a) the balance of 290 acres must be conifer. So the constitution of the final yield for ten years is finally fixed as:

(1) Yield of oak
   = 160 acres (volume about 400,000 hoppus feet)
(2) Yield of conifers
   = 290 acres (volume about 1,350,000 hoppus feet)

TOTAL = 450 acres (volume about 1,750,000 hoppus feet) [1]

[1] In this example yields have been specified by area with approximate equivalent volumes. The yield could have been specified by volume with equivalent approximate areas.

Graphical solution

The problem considered above could have been solved graphically as shown in Fig. 12 and explained below.

Firstly the *desirable* yield is plotted namely the lines denoting 335 acres for conifers and 195 acres for oak. The lines meet at point D.

Secondly the maximum limit for oak is plotted, the vertical line OX, and the minimum limit for conifer, the horizontal line CY, denoting yields of 160 and 230 acres respectively.

Thirdly the maximum limits, irrespective of species, of 470 acres for improving the age-class distribution and 450 acres owing to shortage of resources. As these two limits do not specify species the lines are drawn from the 470 and 450 acre points on the X axis to the 470 and 450 acre points on the Y axis (lines AA and RR in the figure).

Fourthly the minimum area constraint of 400 acres (irrespective of species) for age-class adjustment is drawn (line aa).

From the graph it can be seen that there is a feasible region (shaded) which is contained by the various limiting lines and which therefore satisfies all the constraints. To satisfy the primary object of management, expressed by the desirable yield at D, a yield within the feasible region and as close as possible to D must be chosen. Such a point will lie along the line PQ, any point on which gives a yield of 450 acres.

The point P gives a yield of 160 acres of oak and 290 acres of conifer—the same yield as was calculated previously.

The point Q gives a yield of 115 acres of oak and 335 acres of conifer. Choice between the two points depends on whether it is more profitable to fell and regenerate the oak or the conifer. If the relative profit advantages of felling and regenerating the two species has been calculated profit contours can be drawn on the graph to test which external point on the feasible region is cut. Thus if the profit advantages of felling oak and conifer is as 3 : 2, the contour lines would be as shown in the inset to Fig. 12 as Fig. 12(*a*) (the dotted line through point P is parallel to that drawn from 200 to 300 acres).

That is a simple example partly because the regeneration area, whose selection after a thorough inventory and assessment of the economic status of each stand has been assessed, is positive and clear for a simple prime object, and partly because the

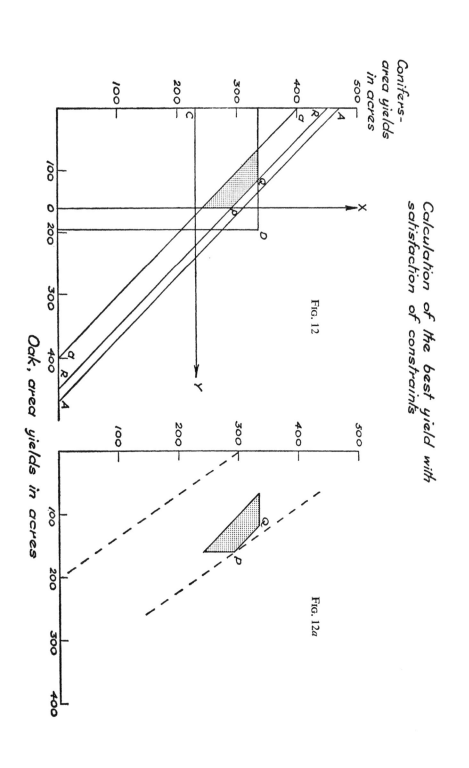

Calculation of the best yield with satisfaction of constraints

Conifers-area yields in acres

Oak, area yields in acres

FIG. 12

FIG. 12a

constraints or variables are few and do not interact unduly either with each other or the desirable economic yield. In practice there might be more constraints and considerable conflict or complications so that solution of the problem might be laborious. For example there might be a need to equalize yearly income which might be affected both by prices for different species and sizes of timber and standing volumes per acre.

When the problem becomes complex, i.e. there is a complicated interaction between choices and there is a number of variables so that a graphical solution is difficult, it is necessary to convert the information to a mathematical model of linear programming (i.e. a programme in which straight lines are involved as in the example that has been considered). The problem can then be submitted to an electronic digital computer for solution.

The subject of mathematical models in forest management has been recently examined at a meeting held at the University of Edinburgh in 1965 and summarized in Forestry Commission Forest Record No. 59 (Makower *et alii*, 1966). The same type of procedure will also apply to the planning of the combined yield from several forests in a region where a desirable or possible yield has to be estimated for the region and then allotted to individual forests each of whose desirable and possible yields must be considered. But in principle the allotment of a regional yield to individual forests does not differ from the allotment of the yield of a forest to individual compartments and sub-compartments. The difference lies in the greater magnitude and dispersal of yields and demands and the greater variety of local conditions so that constraints on prime objects of management become more numerous, various and complex. Correct choice of coefficients for a linear programme becomes more difficult.

III. SUSTAINABLE YIELDS

As essential constraint may be the supply of raw materials to an industry which may expand and may have no good alternative sources for its raw material. The forest or region of forests will thus have to supply not only a guaranteed minimum quantity but a guaranteed sustention of any increase that may become possible so that expansion of the industry is encouraged. There must therefore be no temporary fall of output from the forest below a quantity that is once achieved; a steadily increasing yield without any intervening fluctuations to some final ideal maximum is the object. The normal forest with a sustained yield is the final object, but the stages involved in its attainment must not involve temporary yield reductions.

Methods of regulating the yield are, therefore, similar to those described earlier in this chapter but constrained by the need for the yield never to decrease. To make that possible it is essential to be able to predict the progress of stands to maturity and to know their yields at maturity and during adolescence from thinnings. A thorough reliable inventory is, therefore, required by which all stands are classified by age and yield classes and their volumes and increments ascertained. Yield tables will then make it possible to predict the progress of stands and the yields available in successive periods, of say, ten years. Actual yields can then be equalized by adjusting the cut to husband mature stands for felling later when a shortage of them is foreseen; if younger stands are in excess and if they provide acceptable raw material they can be cut to balance a shortage of old stands.

If suitable yield tables are not available local research must produce them and provide increment and volume data. Without such data and without a reliable inventory of areas by age and yield classes yield regulation will be guess work. Inventories will have to be repeated every period to correct errors, incorporate casualties suffered from accidents of wind, fire or other damage and check yield class assignments of fluctuations in increment owing to cyclic weather changes.

General comments on yield regulation

The several methods of yield calculation that have been described demonstrate methods devised to satisfy the principles which affect yield regulation. The basic principle is to prescribe a cut which guides the existing growing stock closer to that state which best satisfies the objects of management. That constant purpose of improvement is influenced by various constraints imposed by the available resources of finance, machinery and manpower, markets and the state of growing stock and its increment, all of which affect the possible and desirable rate of improvement and the magnitude and distribution of the cut.

There are many other methods and formulae of yield regulation such as those of Hanzlik and the Kootenai formula used
in the U.S.A. (see note below),[1] which observe the same general
principles. Choice of method must depend on the circumstances
prevailing; there is no one best method. Thus the formulae
methods which use increment or the normal forest volume are
useless when the increment or normal forest volume are not
calculable. Where there is lack of a quantitative and silvicultural
knowledge of the growing stock and the influence of sites on
species and growth, yield regulation is inevitably largely guess
work. Such lack of knowledge, comparatively common in
extensively managed tropical forests, is frequently and almost
necessarily combined with doubt on what the objects and policy
of management should be so that there is uncertainty on what
should be grown and silviculturally how it should be grown to
achieve a possibly nebulous purpose. Yield regulation then

[1] (i) The Hanzlik (1922) formula is: $AY = V_m/R + I$, where V_m is the
volume of merchantable timber above rotation age and I is the actual
future MAI of immature timber for the first rotation (R). The quantity I is
accordingly obtained by assuming that it equals the sum of the annual
increment of each age-class (below R) at an age of R/2 greater than it is
now, since R/2 years is the average time that each stand will be left before
it is felled.

The formula is used for the conversion of virgin forest to an even-aged
sustained yield unit. It therefore assumes that the mature untouched
stands have no net increment so that the whole area would be felled in R
years (V_m/R) and regenerated. That rate of felling would be increased in
proportion to the increment of younger stands previously felled and
reconstituted. There is no consideration of thinnings.

(ii) The Kootenai formula (Meyer et alii, 1961) is

$$AY = \frac{A_m}{n} (V_m - V_n) (1 - s)$$

where, A_m = acres of merchantable stands aged R (rotation) or more
years during n years.

n = the period of adjustment.

V_m = the net volume per acre of stands in A_m.

V_n = the planned net residual volume per acre after cutting.

s = a safety factor expressed as a decimal.

The formula is useful in early stages of management when little is known
about the desirable growing stock level; the volume to be left (V_n) after
cutting over a stand is empirically determined on a silvicultural basis.
Thus the trees to be left after cutting might consist of a minimum number
of well-spaced vigorous trees. Such a residual volume method results in
the establishment of a minimum growing stock volume in all worked
stands preparatory to more intensive management.

becomes in reality a matter of combining the utmost use of what is available with the wise conservation of growth potential until more precision of purpose and method can be attained. Satisfactory yield regulation depends on:

(i) Clear objects of management. To avoid the complication and confusion caused by conflicting multiple objects and to follow the precept of not serving two masters, separate working circles with distinct objects of management should be formed. Where multiple objects must be endured in any working circle they should be arranged in a clear order of priority.

(ii) A clear knowledge of the constitution of the growing stock needed to achieve the object. That requires an understanding and separation of the various site-types and an assessment of their potential for growing various species and products and the relation of those products to expected market values.

(iii) A thorough knowledge of the present constitution and increment of the growing stock. The knowledge is not complete without a record of past volumes, increments and yields, preferably by site types which are the same as, or comparable with, those of (ii). Yield tables and production tables are invaluable guides to help assess present even-aged growing stocks and forecast their future performance.

(iv) A sound assessment of markets, present and future so that the products desirable to achieve the objects of management can be compared with what can be grown to form the ideal growing stock of (ii).

(v) A knowledge of the limitations imposed by finance, machinery and manpower or other factors such as communications, previous commitments, rights of common or other legal burdens.

Obviously the more complete the knowledge of the growing stock is—its volume, ages, increment, past development and yields—the more easily can the growing stock be compared with the ideal and its treatment decided. The method of control is an example. Then each working unit (compartment or sub-compartment or stand) can be examined separately and the desirable treatment assessed—whether it be thinning and its intensity or final felling and regeneration—and the consequent yield from each unit computed. That desirable yield must then be adjusted by constraints, such as those of markets or need for a sustained yield, which set upper and lower limits to final yields in particular and perhaps to thinnings or to the global total of yield.

The formulae methods previously described, the best of which is the Austrian or general yield formula, are then useful only to provide the constraint needed for sustention. The crux of the matter (subject to all constraints) is the decision when individual stands should be felled and regenerated by recognition of the moment of ripeness, i.e. the time when a stand fully satisfies the objects of management such as its maximum NDR or peak MAI.

The yield prescribed—by area or volume, global or for final fellings alone as may be—at revision of a plan for a period of years should be flexible so that its detailed fulfilment is left to the local manager (controlled as necessary by his superior) for adaptation to silvicultural variation of stands, market changes and accidental damage. The local manager should yearly forecast the yields he expects in the light of circumstance and within the policy of the plan. The yield prescribed therefore should be a guide to action and serve as an example of the way considered to be best to achieve policy. But divergence from the yield should never be made lightly. For example extra fellings to absorb high prices are dangerous; prices may rise even higher in subsequent years and again encourage excessive fellings and thereby destroy the policy of management.

Yield control

It is wise to keep a continuous watch on what is happening. There should be a strict and constant annual comparison of the yield prescribed by the plan with the yield actually cut and a running total showing the excess or deficit over the prescription to date. This control is particularly important when the interval between revisions of a plan are long.

It should never be forgotten that forests are large, their components are multitudinous and various. Changes in the growing stock as a whole are not easily or quickly or accurately observed visually. Excess or unwise fellings can insidiously deplete a forest particularly one worked under an irregular silvicultural system when the depletion of large trees, scattered as they are over extensive areas, may remain unobserved for several or many years.

Repeated inventories of the growing stock, prescription of a yield (however flexible its application may be) to suit the growing stock and the objects of management and a constant check and record of what is cut with an annual comparison of that actual cut with that which was prescribed are essential features in any forest management.

Bibliography

Amidon, E. L., McConnen, R. J., Navon, D. L. (1966) 'Efficient Development and Use of Forest Land: An Outline of a Proto-type Computer—Oriented System for operational planning', *Forestry Commission Forest Record No 59*. HMSO, London

Anon. (1964) *Manuel pratique d'aménagement*. Ministère de l'Agriculture, Direction générale des Eaux et Forêts

Coker, R. (1966) 'A Brief Introduction to Critical Path Analysis', *Forestry Commission Forest Record No 59*. HMSO, London

Dwight, T. W. (1965) *Confusion Worse Confounded. The Sad Story of Forest Regulation*. Department of Lands and Forests, Ontario

Hanzlik, E. J. (1922) 'Determination of the Annual Cut on a Sustained Basis for Virgin American Forests', *The Journal of Forestry*, XX (6)

Howell, R. S. (1966) 'Simulation by "Monte Carlo" methods', *Forestry Commission Forest Record No 59*. HMSO, London

Jeffers, J. N. R. (1966) 'Nursery planning and production: a preview of a mathematical model of a complex management problem', *Forestry Commission Forest Record No 59*. HMSO, London

Judeich, Frederick (1869) 'The Austrian Cameral valuation method', *Tharandter Forstlicher Jahrberch*, Vol. 19 (1869), translated by T. W. Dwight and P. Jaciev, 1965. Department of Lands and Forests, Ontario

Knuchel, H. (1950) *Planung und Kontrolle in Forstbetrieb*. Verlog H. R. Sauerländer & Co., Aarau

Knuchel, H. (1953) *Planning and Control in the Managed Forest*, translated by Mark L. Anderson. First English Edition: Oliver and Boyd, Edinburgh

Makower, M. S. (1966) 'Introduction to mathematical programming', *Forestry Commission Forest Record No 59*. HMSO, London

Meyer, H. A., Rechnagel, A. B., Stevenson, D. D., Bartoo, R. A. (1961) *Forest Management*, 2nd edition. The Ronald Press Coy, New York

Schlich, W. (1925) *Manual of Forestry, Forest management*, 5th edition. Bradbury, Agnew & Co., Ltd, London

Smythies, E. A. (1933). The safeguarding formula for selection fellings', *The Indian Forester*, 59 (10)

Wardle, P. A. (1966) 'Linear Programming Studies', *Forestry Commission Forest Record No 59*. HMSO, London

Wilson, B. E. (1966) 'Linear Programming as an Aid for Medium Term Production Planning for Paper Mills', *Forestry Commission Forest Record No 59*. HMSO, London

Chapter IX

THE PREPARATION OF
WORKING PLANS

PLANNING IN GENERAL

The satisfactory conduct of any enterprise needs a plan based on a study of all relevant facts to establish firstly the clear purpose of the owning body and secondly the means of achieving the purpose. Without some degree of planning and without control of the execution of an accepted plan, action becomes unco-ordinated, unbalanced and untimely since individuals, influenced by their separate abilities, knowledge and personalities, react differently to immediate events and thereby interfere with each other's acts even if all participants in the enterprise agree continuously with its purpose. The bigger and more complex the enterprise the more necessary is a plan of work which, written and approved after full consideration of all the circumstances and subdivided as necessary for subsidiary operations, defines the purpose of the undertaking and outlines the means for its accomplishment.

Very small undertakings may not require a written plan: the owner or manager can assess the facts and prospects and design a plan mentally and remember it and adjust it to varying circumstances: he can instruct the small manpower verbally and supervise their performance personally. Even then however the owner or manager will undoubtedly keep written notes and records of results to help him control and adjust and improve the plan in the light of experience and as circumstances change.

For continued efficiency an enterprise must be dynamic, never static. An enterprise may be well founded, expertly managed and its operations skilfully devised and controlled so that work has become a smooth routine. But conditions will change, new improved techniques will be invented and perfected. Unless the organization of the enterprise is changed and adapted to incorporate the new techniques and to meet the new conditions, the success of the enterprise will falter until eventually it dies.

It is evident, therefore, that plans must be continuously or at least periodically revised. For revision to be possible and useful there must be a continuous and intelligible record of results and collection of new information on which to base a revised plan. In every plan, therefore, there must be incorporated means for recording results of its activities, for conducting research in its techniques, for collecting information from other sources relevant to its purpose (e.g. population and economic trends) and for digesting the information collected and stored.

The working plan, therefore, should incorporate provision for the succession of processes which permeate a dynamic concern. Those processes contribute to and depend on one another and to a greater or lesser extent occur continuously but nevertheless form a general cycle of events as shown below:

1. Initial collection of facts
2. Digestion and study of the facts and assessment of prospects
3. Choice of purpose
4. Design and approval of a working plan to accomplish the purpose
5. Execution of the plan
6. Record of results, collection of facts by inventory and research

Item (6) replaces the initial item (1) after the first passage of events and the full cycle is then complete and repetitive except that in the second and subsequent cycles the purpose and plan will be revised rather than initiated. In a dynamic and successful undertaking the purpose, plan and consequent activities will expand and increase their scope.

Working plans in forestry

The general principles described above also apply to a forest enterprise so that for all but the smallest forest a written plan is an essential part of management. The working plan enables the owning body firstly to clarify and define the objects of management and the consequent policy of action and secondly to organize the administration and treatment of the forest in harmony with the policy. Furthermore, let us note that firstly, the plan becomes useless and void if it is not performed, therefore control of its execution is necessary; and secondly, wise revision of the plan is impossible unless more facts and full records of its performance are collected. It is therefore essential for the plan to provide for its own control and collection and record of results as essential elements in working the forest.

Essential elements of the treatment of a forest

If the objects or policy of management include a sustained yield there are several aspects of treatment of the forest which become essential and must be incorporated in the working plan. If instead of sustained yields the best sustained production is wanted then the same features of treatment remain essential except for the provision for a steady succession of maturing timber.

Those essential features of treatment are summarized below:

(i) Conservation or improvement of soil and site productivity.

(ii) Attainment of the maximum production of the materials or services desired by the objects of management.

Those two interdependent features imply the need for as skilled and intensive a silviculture as possible in suiting species to site, in the tending of stands and the adjustment or control of stand structure, specific composition and density so that each selected tree gives its best performance without exhausting the site. That is to say, in short, how to cut and tend the growing stock and foster the site.

(iii) The moulding of individual stands and stand types into a composite whole that provides a steady sequence of maturing timber by a planned progress towards an ideal entity.

(iv) The satisfaction of the needs of utilization.

Those two latter principles, combined with (i) and (ii), control the planning of fellings in quantity, place and time—how much, where and when to cut. For sustained production, rather than sustained yields, the emphasis will be more on *when* to cut rather than *how much*, i.e. choice of those stands for felling that have reached the peak of their performance, with forecasts of the quantities of output.

The two principles also affect the planning of sales, transport and communications.

(v) The co-ordination of administrative resources—in men and their skill, money and equipment.

The last feature implies not only that the administration must be planned to suit the treatment but also that the proposed treatment must not exceed the capacity of current resources. The plan must be workable in practice—it must suit demand; progress and expansion must be related to income and capital available; manpower and equipment must be related to progress.

Development stages in management and working plans

The fact that the provisions of a plan cannot neglect the limitations or constraints imposed by the supply of staff, labour and their skill, by resources of capital in money and equipment, by communications and by market prospects, emphasizes two main points:

(i) Changes from previous standards of management cannot be big and drastic—a principle which is also supported by the slow growth of trees.

(ii) There must be a phased progress towards the desired forest constitution, an organic development planned to move in step with improvement of the growing stock, markets and resources in general.

There are of course exceptions which, however, are almost necessarily temporary. For example, if the capital and the will for investment without returns for many years are available a derelict forest could be cleared, reforested and made productive quickly in a few years. But that quick transformation would be succeeded by a long period of waiting as the crop grows and is tended and fostered until its products become saleable and its full management as a growing concern begins.

In that example a period of intense activity has been followed by one of comparative inaction and then a period of slowly increasing action as thinnings are made succeeded by periods of steadily more and more intense management as the forest is brought to full production. Large scale afforestation presents a similar development.

But in general the management of a forest develops more uniformly from comparatively extensive simplicity to intensive complexity.

Let us consider the stages through which the management of a large tropical forest might pass after an initial period of uncontrolled exploitation after which it has been untended, unprotected and more or less devastated in whole or part.

Stage 1. Survey; delineation of boundaries; legal definition of the property. Protection from theft, fire and grazing. Engagement, housing and training of staff. Salvage fellings on a cycle. Sale of firewood and minor produce by permit. Start of a road system. Study of the growing stock and sites.

Stage 2. Formation of compartments and simple working circles for coppice and local supply, selective fellings on a cycle with yield regulation. Stock mapping and inventories. Research in increment and silviculture. Road development. Enlargement and training of staff.

Stage 3. Reduction of size of administrative charges. Redesign of working circles and silvicultural systems to suit sites, forest types and markets. Improvement of regeneration, enrichment and tending methods, of extraction, conversion and transport. Volume and yield table construction. Inventories intensified. Research in market trends.

Stage 4. More reduction of size of administrative charges. Re-adjustment or conversion of systems and species with introduction of fast growing exotics. Concentration on maximum growth and quality to suit markets. Smaller units of working, closer choice of rotation and removal ages. Refined yield regulation. Thinnings a major feature of tending for quality and timing of yields. Records intensified. Attention to detail. Research on site improvement, timber quality.

Naturally the stages are not distinct but merge into one another and do not follow any exact pattern of development from extensive simplicity to detailed intensity. Naturally also the speed of progress varies with conditions such as the original state of the forest, size and accessibility of the area and growth rates of the growing stock and markets. But whatever the pattern and rate of development may be the current stage depends on past stages and the working plan which directs the current stage, depends on the past plan and influences the next. Each working plan is a link in a chain of practical experiments using the results of past experiments and solutions of past problems in the light of current conditions to lead the forest to a more desirable state.

The intensity of working

The intensity of a plan, i.e. its thoroughness and attention to detailed treatment, naturally varies with circumstances and the stage of development already reached. In general as the income from a forest increases, especially if treatment is necessarily complex owing to variety of site, species and also purpose, so does the intensity of management (and of the plan) increase. The whole *tempo* changes justifying greater attention to detail to improve production still more without unnecessary cost.

Factors that affect intensity of working are:

(i) *State of demand and financial prospects.* This is influenced by several matters, mainly:

(*a*) Accessibility which depends on topography and distance from population centres.

(*b*) Value of the products of the growing stock and their quantity.

(*c*) The markets available which depend on (*a*) and (*b*) above and on growth of the population and standards of living.

(ii) *Stage of forest development reached.* The stage of development is shown by several factors:

(*a*) The state of the growing stock, in particular by its current state of productivity, variation in species and site and its potential capacity.

(*b*) The state of silvicultural knowledge.

(*c*) The state of communications in and outside the forest, by road, rail or water.

(*d*) Previous plans and the degree of intensity reached in them. Detailed and intensive working can seldom or never be achieved without preliminary phases of development.

(iii) *Resources available.* The most compelling resources are the skill and numbers of staff and the capital available in money and equipment. The extent of these depend essentially on items (i) and (ii) above.

Greater intensity of working, namely attention to smaller units of land and stand and to detailed treatment of stands (e.g. pruning), does not necessarily involve an equivalent complexity of work. Intensive plans can be comparatively simple to execute provided that standard practices can be used so that their execution can become a routine.

In any event a sound principle is that both the treatment of a forest and the plan prescribing treatment should be as simple as circumstances allow and no more detailed than can be justified by the cost/advantage ratio.

Long and short-term planning

To obtain a phased progress in the management of a forest each successive plan must provide for both the long and short-term.

The long-term planning endeavours to secure continuity of both purpose and the consequent policy of main method. Some subsequent changes will be inevitable but they should be minor, those of degree and rate of progress and not those of purpose, policy and basic method. So the general objects of management and policy which must include decisions on the type of product wanted must be soundly conceived so that they remain stable. The consequent silvicultural method to give a growing stock that fulfils the purpose must also be soundly conceived; for example the species concerned and whether they should be grown in an even-aged or uneven-aged system. Together the objects and basic silvicultural methods settle the allotment of working circles which should at the same time be divided as necessary into felling series for market, administrative or silvicultural convenience. The fundamental and long-term organization of the forest will then have been settled; subsequent change from it will be difficult, long and expensive.

Additionally the long-term plan should outline the main phases or stages of treatment that are expected to be necessary for the development of the forest as a whole and of each working circle in particular. For example there may at first be a stage of protection with salvage felling and road development, succeeded by a phase of conversion to even-aged plantations of exotics, succeeded by a phase of thinning and adjustment to normality and intense development of markets.

The long-term planning by determining the general policy and basic methods to be employed forms the foundation on which management and subsidiary, detailed planning rests. Changes in this fundamental design should be made with reluctance as it should, if soundly conceived, be long-lasting and need little or no adjustment at each revision of the plans.

The short-term plan is concerned with what is to be done next, i.e. during the next phase of the long-term plan, the next link in the chain of management. It will therefore prescribe action for the next ten years or whatever may be the period of the plan, regulating how much should be cut and where, when and how it should be cut and also controlling the manifold other activities of protection, road repair and construction, sale methods, records and so on to the extent that is necessary beyond standard practices that have been embodied in general instructions and procedure.

The working plan area

The extent of a working plan area (synonym in the USA is working circle and in Canada sustained yield unit), i.e. the area included in the provisions of one plan, depends on administrative convenience. Accordingly it should ordinarily coincide with an executive charge or with one ownership if an executive charge includes properties of more than one owner since separate owners are unlikely to accept the provisions of a common plan unless they have agreed to close co-operation in the management and use of their forest properties.

But executive charges vary in type, for example that of the beat or range of the forester and that of the district or division containing several beats or ranges, in charge of a district or divisional officer. Whether the working plan area coincides with the larger or smaller unit depends on two main factors—the degrees of dispersion and development of the forest.

In a large forest in an extensive, early stage of development it is often convenient for the working plan area to coincide with the range or beat which itself is likely to be comparatively large. In such forests the degree of development is likely to vary considerably from range to range owing to differing states of accessibility and demand. So plans are first made for the more accessible, more heavily worked range after which plans are made for the other ranges as they are developed in their turn. As management intensifies in course of time the several plans can be blended to form one plan for the whole district.

Similarly a district officer may be in charge of a highly dispersed forest whose several ranges or beats are distinct and widely separated. Separate plans for each range or beat may then be convenient particularly if the types of forest and market are also distinct. Even then one plan may be the more convenient if it contains several working circles or felling series, i.e. subsidiary plans coincident with the ranges or beats. Subsequent development and intensification of management may later necessitate raising the original large ranges to the status of districts; the several working circles or felling series can then easily form distinct working plan areas.

Working plan areas, therefore, tend to change with the stage of forest development and intensification of work, demand and yield, modified by the degree of dispersion of the forest.

In the USA and Canada the working plan area coincides more with a forest which is a sustained yield unit and is often related to particular forest industry or consortium of industries so that the area is established in terms of production and consumption of those industries (H. A. Meyer *et alii*, 1961).

In Britain there is now a tendency to consider the adoption of regional working plans so that the working plan area would include a region in which there are several forests and district officers' charges. The region would be a sustained yield unit although the several forests might have fluctuating yields based on fellings which aim at attaining maximum sustained production. A working plan area of that type would make it unnecessary for local forests to have a sustained yield. Instead the sustained yield unit becomes a region, perhaps 100 miles in diameter or some 7500 square miles in area, whose 200 square miles of forests might serve a more or less centrally placed industrial community.

It would seem that the scope of regional working plans would be only to direct policy and co-ordinate output from the component forests each of which would still have its own plan to control treatment which would conform with the regional policy so that supplies of raw material and services suit the industrial development and life of the region.

The scope of working plans

A working plan should not merely express and provide long and short-term planning. It should also summarize the reasoning that justifies the planning and proposed action. Without records of such reasoning, subsequent critical analysis of the plan when considering changes and progress for the next plan will be handicapped. That reasoning, however, depends on the digestion of past and current facts and their interaction and on the forecast of future events. The reasoning therefore requires an assessment of the growing stock and sites where it grows, of the results of past treatment and an assessment of current demands for raw material and services and an appreciation of the trend of demand. So the plan must include a stock-taking, a survey of resources and a study of limitations which together are analysed to enable decisions on policy and basic methods and detailed action to be made.

Furthermore the plan must also provide for the maintenance and digestion of reliable records of progress and for research and collection of new data so that the next plan can be based on still better facts.

So the broad scope of a plan is fourfold:

(i) A survey and assessment of past results, present facts, resources and limitations.
(ii) An analysis of those facts with reference to prospects leading to conclusions on policy, objects of management and basic methods proposed—the long-term plan.
(iii) A plan for future action in a definite period in conformity with the long-term plan—the short-term plan.
(iv) Provision for control of the prescriptions, for maintenance of records and collection of new facts by research.

In that summary of the broad scope of working plans the omission of financial considerations and forecasts of material yields and money income does not imply that they have a small place in plans. On the contrary they must play a great part and permeate many sections of the plan, particularly the long and short-term plans and the assessment of resources and their potential. A special chapter or section may also be needed to discuss the yields and net income expected in the plan period and beyond.

The working plan period

Clearly no plan can be made once and for all. Circumstances always change; markets will vary with changing accessibility, population, standards of living and other factors. Accidents of fire, wind and the like occur or the growing stock may not develop as expected nor match the results obtainable from other species and new techniques.

So plans are made for a limited period. The length of the period will depend on the stability of markets, methods of treating the growth stock and the degree of its response. If the forest is experiencing the protection stage of rehabilitation or the period of adolescence from afforestation when demands are small, the period may be long such as twenty years. Conversely for an intensely managed forest, affected by changes in industrial demand and new knowledge and new techniques obtained from active research, short periods of ten years or even five years are advisable especially if the species used grow fast.

It is convenient for the revision of plans to coincide with definite events in the management programme such as the end of successive regeneration periods. That has long been the practice on the continent of Europe. As regeneration periods may be comparatively long, e.g. twenty or more years, it is then advisable to repeat inventories and re-assess the yields more frequently such as at half time.

Plans should not be revised too soon, in too much of a hurry. Not only is revision costly but sufficient time must be given to prove the success or failure of prescriptions and test the permanance and reliability of promising new markets or methods. In Britain the period of plans is usually ten years although five years is the present period for private forests managed and developed with the aid of monetary State grants.

It may become essential to revise a plan before it expires on the occurrence of major unforeseen events such as a calamitous fire, a great economic depression, war, failure of an industry on which the forest relies or new accessibility caused by a national assessment and reconstruction of roads and transport.

Annual plans of operations

Although working plans summarize treatment for a period of some ten years there still remains the need for yearly planning of programmes of work within the framework of the plan. Accidents may happen, changes in demand, labour supplies or money available may occur which necessitate the adaptation of even rigid, detailed prescriptions of a plan. Moreover many plans are not rigid but merely outline the principles on which felling, road construction and other work are to be based leaving details of their execution to the manager.

Annual plans of operations are, therefore, necessary in which are incorporated the details of work, in quantity and locality, for the year. This plan of operations supports and agrees with the budget allotment of money and its distribution to various types of work, and forecasts the income expected from the yield which is distributed among fellings that conform with the limits prescribed in the plan and the demand expected.

Annual plans of operations also confirm and allot programmes of work to the respective sub-managers, the foresters or rangers in charge of beats or ranges, so that they have a clear knowledge of their responsibilities for the year. Forecasts of yield made in annual plans of operations also make it possible for higher authority to adjust deliveries from several forests to industry or for industry to adjust their arrangements for supplies of raw material.

Authority

Until a plan is accepted and translated to action it remains a recommendation only and the execution and accomplishment of a plan requires the approval and authority of the owner or his representative who then can enforce its execution and allow deviation from it. The bigger and more complicated the enterprise and governing body are, the more careful must the process of scrutiny of the recommendations be before they are approved and put into action as a plan of work. In a large, complicated concern there will necessarily be much delegation of authority and responsibility in making and co-ordinating subsidiary plans whereas in a small affair the owner can personally attend not only to policy and main methods but also to detailed techniques and daily work. Very careful scrutiny and check of plans for large concerns before their approval is obviously essential.

Forest working plans are no exception particularly those for State forests whose management involves not only substantial amounts of money in capital, income and current expenditure but also affects the well-being of many people and industries so that the working and output of individual forests influence and are influenced by those of other forests. The management of each must conform with that of the others and with regional forest policy. Moreover, the compilation of a plan requires much skill and judgment in the assessment of the productive capacity of the forest and its sites, in deciding the purpose of management and in choosing the means of accomplishment. Decisions made are often controversial and yet, as mistakes are not easily or quickly rectified, are bound to affect results for a long time. There must therefore be close collaboration between the compiler of the plan and those who will apply it, and also a system of critical analysis and check of the provisions of the plan and its final approval by a high authority.

The need for working plans to be sanctioned by a high authority, after recommendations from a chain of lower authorities, which may include a specialist working plan branch, has long been recognized. Working plans in France once had to be sanctioned by the President of France; Indian working plans require Governmental sanction. Such high authority for a plan *in toto* makes it difficult to obtain permission to adjust the plan in any way subsequently, until it expires. In general and in principle that is a good thing since it prevents hasty and subjective changes without critical consideration of relevant facts, changes which would tend to defeat the object of having a plan at all. But small changes which do not interfere with policy and main methods but merely adjust minor matters in the light of experience may be eminently desirable. Therefore, if the plan *in toto* is approved by a high authority it is necessary to include in the plan provisions defining the types of deviation that can be approved by lower authorities without reference to the higher. Alternatively a system could be adopted whereby only the policy and basic methods are sanctioned by the high authority and the means of executing that basic planning are sanctioned by a lower authority.

The latter procedure has been adopted by the Forestry Commission for State forests in Britain. When a plan is compiled a summary and analysis of the relevant facts and prospects of the forest are first made together with reasoned recommendations for the policy and main methods of management that are proposed, i.e. the long-term plan and the facts and reasoning in support of it. These are sent to the Director-General, the head of the State Forest Service and, as necessary, to the members of the board of Forestry Commissioners (of which he is a member) for approval. He can therefore verify whether the policy and basic methods proposed for the forest are sound and conform with those of the region and national policy. After his approval, the short-term plan, i.e. the ways and means of accomplishing the long-term plan during the working plan period, is written under the direction and subject to the approval of the Conservator, the next lower authority. The plan should then indicate to what extent and what type of alterations can be made by the local manager or district officer without reference to the Conservator.

Control forms

To ensure proper execution of the provisions of a plan, a form of control, which records what is actually done with what ought to have been done, is necessary. Deviations from the prescriptions would require explanation and approval by the appropriate authority. (For an example of a control form see page 286.)

A preliminary working plan report

Before the work of preparing a plan is begun it is often necessary for a preliminary working plan report to be prepared. This preliminary report is particularly necessary before a first working plan is made. The report is based on a quick, general survey or reconnaissance of the area, the growing stock, market prospects, previous treatment and the general constraints that affect future treatment. A summary and analysis of the survey is embodied in the report which includes the author's opinions on the need for a working plan or its revision and his recommendations on what the purpose of management and the general long-term policy should be, including the main managerial and silvicultural methods needed and a short summary of the short-term prescriptions contemplated for the next stage of work. Additionally the report should include an estimate of the work that will be involved in preparing the plan, the manpower and the time needed and the cost.

The controlling authority then appoints a working plan officer, staff and money for the job with instructions which embody a provisional decision on objects of management, on the long-term policy and main methods of treatment that will probably be adopted. These provisional decisions, though liable to be modified when field and other work have collected all the facts, are necessary as they may influence the type and intensity of the work of preparing the plan—e.g. the intensity and consequent organization and methods of the inventory.

A comprehensive preliminary working plan report becomes less necessary at revisions of an existing plan, particularly if policy and methods have become stabilized unless some catastrophe such as a disastrous fire, a major and lasting trade slump or a war upsets the stability. But even when the revision of a plan becomes a matter of routine it is valuable, even essential, for the manager in charge of the forest to write an appreciation of the results of current working, an appreciation which analyses the success and imperfections of methods, organization and techniques, the development of markets and so forth.

The important thing is that due consultation and consideration by the controlling authority should precede the preparation of the actual plan or its revision so that the work of preparation can be efficiently organized for an understood purpose and main methods of accomplishment. Much waste of effort, time and money may then be avoided.

The arrangement of a plan

Although no universal standard is possible or desirable it is clearly advantageous for the contents of a plan to be arranged in a natural, logical sequence so that the need for forward references and repetitions is avoided. It is also desirable that the main arrangement of plans within a country should be standardized as far as local circumstances allow so that all members of the professional staff, e.g. of a forestry service, are familiar with it and its mode of execution.

Accordingly State forest departments usually have working plan codes which outline working plan procedures and list the headings of the contents of the working plan to serve as a basic guide in the compilation of plans. Examples are the British Working Plans Code (1965) and the several early working plan codes for the individual provinces (now States) of India, e.g. that for Bihar in 1927. Several authors have also proposed formats for working plans, e.g. N. V. Brasnett (1955) who adopted and changed slightly that proposed by R. Bourne (1934) which was based on Indian experiences, and H. A. Meyer *et alii* (1961) for forest management in the USA.

For the purpose of this book, in order to indicate an arrangement that may provide principles suitable in many countries and to provide a convenient basis for comment, a fairly comprehensive format is given below. It will be seen that the format is composed of five main parts, preceded by a foreword, a table of contents, summary of the plan and an introduction. The five main parts are:

> *Part I*   Survey of the facts
>   A. The estate and basic site factors;
>   B. The growing stock;
>   C. The administration and economy.
>
> *Part II* Foundations of the plan—analysis of the facts, settlement of purpose and consequent main methods; division of the area
>
> *Part III* Management of the growing stock for the plan period:
>   A. ⎫
>   B. ⎬ Subsidiary plans for the several working
>   C. ⎭ circles
>   etc.,
>
> *Part IV* General management of the estate
>
> *Part V*   Appendices and supporting maps.

GENERAL FORM OF A WORKING PLAN

TITLE PAGE
Name of forest and its owner. Period of plan. Author. Date.

FOREWORD
Authoritative sanction for the plan with relevant correspondence.

TABLE OF CONTENTS

SHORT SUMMARY OF THE PLAN

Period of plan.
Area, by land uses.
Objects of management and main methods to be used. Division of the area
    by working circles and felling series.
Main prescriptions for the growing stock and forecasted annual yields.
Main prescriptions for the estate as a whole.
Forecasted income, expenditure and net balance.

INTRODUCTION

Reasons for preparation of the plan with reference to preliminary reports.
Organization and timetable of the work of preparing the plan.
Cost of the plan.

THE PLAN

Part I. Survey of the Facts

A. The estate and basic site factors

CHAPTER I.   Situation, extent, history and legal status of the estate

    Section (*a*) Name, situation and owner
            (*b*) Boundaries
            (*c*) Area by land uses
            (*d*) General history of the property
            (*e*) Legal status of the property and commitments

CHAPTER II.   Permanent site factors

    Section (*a*) Topography, altitudes and drainage system
            (*b*) Climate
            (*c*) Geology
            (*d*) Soils and soil type classification

CHAPTER III.   Variable site factors

    Section (*a*) Man
               (i) Theft, trespass and interference
               (ii) Domestic animals
               (iii) Fire
            (*b*) Wild vertebrates
            (*c*) Invertebrates
            (*d*) Fungi and other vegetative interference

CHAPTER IV.   The ecology

    Section (*a*) Floral associations and faunal influence
            (*b*) Site type analysis and classification
            (*c*) Suitability of tree species to sites

CHAPTER V.   Artificial structures
    Section (*a*) Roads, bridges and rides
          (*b*) Buildings
          (*c*) Drainage
          (*d*) Fences and other structures

B. The growing stock

CHAPTER I.   Silvicultural assessment
    Section (*a*) Past silvicultural treatment
          (*b*) Present forest types—specific composition and structure, health and vigour, extent and distribution
          (c) Current methods of treatment

CHAPTER II.   Quantative assessment
    Section (*a*) Methods of stock mapping and mensuration used
          (*b*) Summary of inventory—by territorial or management units or forest types as convenient
              (i) Areas (or frequencies) of age, treatment or size-classes
              (ii) Volumes of classes
              (iii) Increment of classes
          (*c*) Summary of past records of intermediate and final yields by chosen units

CHAPTER III.   Management assessment
    Section (*a*) Summary of past working plans
          (*b*) Present silvicultural management analysed in relation to:
              (i) Current objects of management
               (ii) Proportional distribution of age, treatment or size-classes
              (iii) Relation of increments to yields
          (*c*) Potential for improvement, such as suitability of
              (i) Species
              (ii) Rotations or removal ages
              (iii) Silvicultural techniques and methods

C. The administraton and economy

CHAPTER I.   The administration
    Section (*a*) The administrative organization
              (i) Managerial and sub-managerial charges
              (ii) Technical and clerical establishments
              (iii) Pay scales
          (b) Man-power
              (i) Sources of skilled and unskilled labour
              (ii) Housing and wage rates
          (*c*) Suitability of organization

CHAPTER II. Forest services

Section (a) Types and extent of demand
        (b) Means of satisfaction

CHAPTER III. Material forest products

Section (a) Markets and extent of demand
        (b) Methods of sale
        (c) Methods of conversion, extraction and transport
        (d) Summary of quantities and assortments of produce supplied

CHAPTER IV. The economy

Section (a) Prices for forest produce
        (b) Costs of harvesting—felling, conversion and extraction
        (c) Costs of regeneration, tending, protection and maintenance
        (d) Cost of supplying services
        (e) Taxation and subsidies
        (f) Overhead charges
        (g) Profit and loss account and balance sheet

Part II. The Foundation of the Plan

CHAPTER I. The present position and prospects

Section (a) The trend of demands, their relative values and mutual compatibility
        (b) Suitability of growing stock for the demand—species, structural type, health and age range
        (c) Suitability of current organization
            (i) Silvicultural systems and working circles, rotations and tending
            (ii) Roads and transport
            (iii) Administration and economy
        (d) Influence of current commitments and other constraints

CHAPTER II. The long-term policy

Section (a) Analysis of results of past policy and methods and possible alternatives
        (b) Choice of management with priorities and regions of application
        (c) Influence of objects on products to be grown—types of species, range of sizes or ages

9

CHAPTER III. The main methods chosen

Section (a) The types of ideal growing stock desired related to main objects and localities of application
        (b) Consequent silvicultural systems and conversion methods selected
        (c) The units of working to be used
        (d) The administration and system of communication

CHAPTER IV. The framework of the plan

Section (a) Division of the area
                (i) Administrative charges
                (ii) Compartation
                (iii) Working circles and felling series
        (b) Period of the plan

Part III. Management of the Growing Stock in the Period of the Plan

A. The working circle

CHAPTER I. Management design

Section (a) Particular objects of management by priorities
        (b) The products to be grown and standards of maturity
        (c) Situation, extent and sub-division of the working circle

CHAPTER II. Silvicultural design

Section (a) Summary of the state of the growing stock—by age or size-classes, their areas, volumes and increments
        (b) Peculiarities of growing stock in relation to objects and summary of treatment proposed
        (c) Choice of rotation, regeneration period and block or area, felling cycle and principles for removal ages

CHAPTER III. Regulation of the yield

Section (a) Analysis of relevant factors
        (b) Calculation and prescription of the yield
        (c) Allocation of the yield
                (i) The final felling plan with relevant directives
                (ii) The thinning plan with instructions

CHAPTER IV. Regeneration and early tending

Section (a) Ground preparation, drainage and fencing
        (b) Planting and plant requirements
        (c) Beating, weeding and cleaning
        (d) Brashing and pruning

CHAPTER V. Tending, harvesting and yield of minor produce

CHAPTER VI. Forecasts

Section (a) Forecasts of annual outturn
            (i) Final yields
            (ii) Thinning yields
            (iii) Minor produce
       (b) Comparison of growing stock at beginning and end period

Part IV. General Management of the Estate

CHAPTER I. Protection

Section (a) Protection from fire
       (b) Protection from man and domestic animals
       (c) Other protection

CHAPTER II. Plant supplies and nursery operations

CHAPTER III. Exploitation and economic development

Section (a) Felling and extraction
       (b) Conversion and sawmilling
       (c) Methods of sale
       (d) Development of markets and forest industries

CHAPTER IV. Permanent structures

Section (a) Buildings
       (b) Roads, bridges and communications
       (c) Main drainage
       (d) Other structures

CHAPTER V. Administration

Section (a) Grazing, sporting and other lettings
       (b) Rights and concessions
       (c) Administrative and supervisory staff
       (d) Labour
       (e) Tools and equipment
       (f) Research

CHAPTER VI. Controls and records

Section (a) Control of prescriptions and control forms
       (b) Compartment histories and other records
       (c) Records of fellings, inventories and surveys

CHAPTER VII. Finance

    Section (*a*) Summary of income, expenditure and capital value of
              ancillary enterprises
          (*b*) Capital value of the estate now, and forecasted value
              at the end of the period
          (*c*) Financial forecast
              (i) Profit and loss account
              (ii) Balance sheet
              (iii) Analysis of financial position and prospects

Part V. Appendices

1. Glossaries of botanical and other terms
2. Detailed area statement—by compartments for territorial and management units
3. Results of inventory
      (i) Areas of age-classes or frequencies of size-classes, by species, working units, compartments or other unit as required
      (ii) Volumes of classes
4. Examples of calculation methods
5. Description of compartments
6. Maps, such as:
      (*a*) Reference map
      (*b*) Geology and soil types
      (*c*) Vegetation types
      (*d*) Site types
      (*e*) Growing stock
      (*f*) Management

PART I. COMMENTS ON THE SUBJECT MATTER OF A PLAN

Essentially Part I is a collection of facts which should be collected continuously as a part of ordinary management to form a permanent record which is then sifted, supplemented, brought up to date and summarized at each revision of the plan. Some of the information may change little or not at all with the passage of time, e.g. permanent site factors such as the geology, but may have to be expanded as a result of new knowledge. Other information may change substantially as the result of treatment, such as the constitution of the growing stock or, as the result of changes of population or living standards and their consequent effect on markets. Previous information may then have little but historical value except to indicate faults to be corrected or to show a trend that is developing and thereby facilitate forecasts of future prospects.

It is useful, therefore, to keep permanent files in which information can be recorded as it occurs or is discovered for record under the same main headings as those used in the working plan.

As it is essential for a working plan to be as short and concise as possible without excluding necessary information, the extensive use of tabular statements, histograms and diagrams which can eliminate or shorten lengthy explanations is desirable. Conclusions should be justified by short reasoning or figures or, when possible, by reference to published opinions, to departmental correspondence and research results or to accepted practice. Nevertheless the plan should be readable and form a coherent whole which preferably can be understood by persons other than the local staff. This need for concision applies to Part I as well as all other parts of the plan although being the descriptive part, some degree of latitude is acceptable; it is written for reference and not daily use.

It is worth noting that negative information in Part I can have both current and even greater historical value. Examples are 'no sign of rabbits was seen' or 'no invasive weed such as bracken is present'.

We can also note that although Part I should indicate and emphasize successes, failures and omissions of treatment it can only suggest what means of improvement are possible without any prescription for action. The actual prescription can only be made later in Parts II, III or IV after all matters of resources and limitations and their relation to the possible improvement have been considered.

A. The estate and basic site factors

Situation, extent, history and legal status

This chapter seeks to present the basic attributes of the estate and its legal foundations and limits. Matters of importance are:

Situation. A small scale map is useful to support description and to show the communications and distances from towns and markets.

Boundaries. Only a short description of the types of boundaries used and a summary of their state of repair are necessary with reference to the documents which establish legal recognition of the boundaries.

Area and land use. A tabular statement such as in Table 23 which classifies the land use is necessary, together with explanatory remarks and indication of peculiarities.

Table 23

CLASSIFICATION OF LAND AREA BY USE

(areas in acres)

Main Forest Unit Blocks etc.	Inaccessible	Forest Land					Total Forest	Other Land			Total Land
			Accessible					a	b	Total	
		Productive	Unproductive		Total						
			Usable	Unusable							
TOTAL											

Explanatory notes must define the meaning of accessible and inaccessible, productive, unproductive, usable and unusable. Thus an area felled but only temporarily bare of trees is productive and usable; a marsh would be unproductive but if reasonably capable of drainage is usable. Inaccessible forest would be that beyond the present reach of useful communications whether the forest is potentially productive or not.

Rides, narrow forest roads and streams are usually considered to be part of the productive forest; main roads and waterways would be classed as 'Other land' under one of the several heads shown as a and b. The extent to which 'Other land' such as residential or office sites, agricultural land, sawmills and the like are included in the working plan area will depend on circumstances; they are usually excluded.

General history. The object of this section is to lead up to and clarify the present legal position and administrative authority summarized in the next section. Therefore it is not a silvicultural history but rather one of past methods of land use, vicissitudes of ownership and development of rights of usage, of legal enactments and court orders which have affected the use of land or the confinement of particular use of rights to particular parts of the area, e.g. by enclosures.

Legal status and commitments. This section summarizes the legal position, e.g. freehold, leasehold, etc., and states and defines the rights of usage that burden all or parts of the estate and enumerates the privileges that may be exercised. Relevant enactments and gazette notifications (e.g. of reserves or enclosures) and legal documents or deeds of ownership and restrictions on it should be quoted and summarized as necessary. A tabular statement by woods, enclosures or reserves (possibly in an appendix) should summarize the extent of freeholds, leaseholds and rights. In addition a statement should summarize the extent of current commitments such as leases for exploitation of timber or agreements for supply of water or minerals.

A final statement of ownership should describe the administering authority and responsible agents for the owner. Thus a State forest may be owned by the people to whom an elected minister is responsible; the minister delegates administrative and executive authority to the State Forest Service to act within the policy decided by the government.

Permanent site factors

The permanent site factors control the forest products that it is possible to grow and their rates of growth. They also influence the purpose and methods of management. The account of those factors, therefore, should endeavour to stress those features which are critical to growth limits, purpose and methods.

Topography and altitude. The account should summarize the general configuration of the land, the range of altitudes and the sharpness or absence of relief with comments on their effect on tree growth, land stability and forestry practice such as extraction of timber. Exposure to high or salt-laden winds, the prevalence of frost hollows and drainage difficulties in a flat terrain are some obvious examples. Local variations and exceptions may require emphasis. The description should not omit references to the main waterways and the natural drainage system.

Climate. The description of climate should be supported by metereological data recorded wherever possible within the forest itself. Data maintained for many years are particularly valuable as they may indicate climatic cycles which may help the interpretation of variations in increment.

Climatic features which have particular importance are the extent and distribution of rainfall, especially the distribution during the growing season, the prevalence and times of droughts particularly if they coincide with high temperatures and low humidities. Low temperatures may also be highly critical not so much when they occur in seasons of vegetative rest in temperate regions but more when they occur as late frosts at the beginning of the growing season or as early frosts before late growth has had time to 'harden off'. In tropical or sub-tropical regions frost at any time may be decisive so that elevation and frost hollows may control closely the distribution of many species. High winds may be another serious factor especially if they coincide with wetness and tree canopies in full leaf or with the season of early growth when young, rapidly growing shoots are easily blown off or broken.

The discussion should relate the topography closely with the critical climatic features whose intensity may vary greatly with the local configuration, aspect and relief.

Another important factor is the occurrence of dryness (possibly in windy conditions) during the dormant vegetative season when grass and other herbaceous vegetation are composed of dead, combustible material and the risk of severe forest fires is very high.

Geology. The geology has several fundamental influences particularly if the soils are *in situ*, derived from the rocks below. A geological map which includes vertical profiles to show the sequence and distribution of rock formations should support the description which summarizes the results of a geological survey.

Important features are the dip and strike of the rocks, their nearness to the surface, porosity, structure and tendency to fracture or disintegrate to provide channels for water percolation and penetration. Their chemical composition will also greatly influence the fertility of the soils derived from them.

The section should be written on the one hand to support and interpret the configuration and natural drainage of the land and on the other hand to help interpret the composition and properties of the soils to be described in the next section.

Soils and soil classification. This section should summarize and implement a pedalogical survey to which reference can be made for details. The extent and detail of such a survey will naturally depend on the intensity and degree of development that the management has attained and the consequent need for a detailed or broad classification of soil types.

The classification adopted should be a practical one by which its several elements are comparatively easily distinguished in the field and have definite effects on vegetative growth. Usually, therefore, the classification will depend on soil depth, texture, structure, water regime and the degree of alkalinity or acidity interpreted by the hydrogen ion content or pH value. Colour is also likely to be a valuable aid in studies of the soil profile to show the state of development between a brown forest soil and a podsol, to show relative degrees of oxidation or reduction at varying depths and the decomposition, accumulation or re-deposit of organic or other material.

It is useful for the description and classification to be illustrated by coloured diagrams or representations of the profiles of each soil type with a list of the vegetation that accompanies it.

In studying and assessing the soils it must always be remembered that the final object is to recognize, classify and map site-types to which definite forest types, species and possibly also treatments can be assigned. The site-types will depend not only on soil but on other factors too such as the local configuration of the land and consequent drainage and local climate which, admittedly, also affect soil. But soils may also be affected by the vegetation they carry. That should not be forgotten in the soil study not only because the vegetation and past treatment of the land may help to interpret the character of the soil but also because it may prove undesirable to inflict a soil with an otherwise desirable species and treatment. Such consideration in compiling a soil-type classification may influence the subsequent site-type classification.

Variable site factors

This chapter should be written from two main aspects, firstly that the factors may greatly influence the growing stock and forest operations and secondly that they are controllable to a greater or lesser degree. Therefore, on the one hand the account should assess the prevalence of the factors and the amount of harm (or good) that they cause and on the other hand it should describe the measures so far taken to combat the harm (or encourage the good) and assess the degree of success achieved with an analysis of the reasons for any imperfection. Costs of measures taken should not be overlooked.

Negative information—for example the absence of an animal or vegetative pest, such as Loranthus or a species of climber, may prove very valuable subsequently if the pest or benefactor were to appear or disappear after methods of treatment or protection have been changed.

The account of vegetative interference should not unduly overlap the ecological assessment made in the next chapter. Some overlap may be inevitable, particularly in intricate and profuse communities such as in moist evergreen tropical forest, if control measures are linked with the general ecology.

The ecology

The object of a study of the vegetation and its ecology (and of the fauna as far as may be necessary) is to promote an understanding of the most satisfactory means of growing trees or other forest produce in various parts of the forest—in short to indicate what to grow and how to tend the growing of it.

Although the extent and details of information wanted from an ecological survey will naturally vary with circumstance and local problems, in broad terms the information wanted embraces:

(i) A map showing the distribution of plant communities or vegetative types.

(ii) The character of each of the vegetative types and the parts played by its main components.

(iii) The relation of each vegetative type to site conditions (soil, land configuration, water regime) to past treatment and to other vegetative types (e.g. the seral relations).

(iv) A classification and map of site-types with a description of their properties.

(v) The character of selected species (particularly trees) and their performance (including regeneration) in different sites and their effect on each other.

(vi) The response of sites to silvicultural treatment, plant communities and main species, particularly to pure tree species and invasive herbaceous or shrubby species.

(vii) The need for and prospects of the introduction of new tree species.

(viii) The response of sites to physical treatment such as draining, irrigation and fertilizers.

The working plan cannot be expected to collect full information on that vast field of knowledge or to give more than a summary of the information available on the salient problems to be solved with references to other sources. The point being made here is that the ecological assessment made in a working plan should be eminently practical to show where the productive capacity of the land, site by site, varies and how each site can be used and improved beyond the state of development that has already been achieved. The account and conclusions must be linked with the problems being encountered, problems which have an infinite variety, from the difficulty of increasing the useful stocking of dry, tropical savannah to the planting of heather-clad, temperate moorland with forest trees. In the former fire-protection and choice of species may be basic, in the latter ground preparation, drainage and choice of species. In both the effects of past treatment, the extent of soil deficiencies and the distribution of plant communities will help in the choice of species, in the methods of tending them, and in deciding the means of improving site conditions.

In many forests the desirable degree of mixture and the best type of vertical structure of stands are debated problems. In Europe, for example, to what extent should a species of comparatively low value or productivity, such as beech, be grown as a silvicultural subsidiary or understorey to the main species? In evergreen, tropical forest should the multiple specific composition and multifold vertical structure be reduced to a select band of valuable species in even-aged stands? The modern tendency is to use a selected species, frequently an exotic, pure and even-aged with artificial regeneration. Is that wise in the long term? Is it probable that treatment of the site—by drainage or irrigation and wide use of artificial fertilizers—will sustain site productivity and harmony of the desired species with the site, and also prove profitable? Or is it better to use the local indigenous species and the plant community associated with it even if its production is less than that now promised by an exotic?

Those are basic problems of forest management which can neither afford to grow less desirable material however fast growing or however preservative of the site the species may be, nor allow sites to deteriorate consequent to the use of a more commercially desirable species. The ecological survey must help management by providing evidence to delineate and assess sites, to suggest the most suitable species for the site in the prevailing circumstances and to judge the needs of the species and site so that the best treatment for each can be decided.

B. The growing stock

The management of a forest cannot be satisfactorily decided or accomplished without a complete knowledge of the present state of its growing stock, the reasons for that present state and its suitability for the purpose of the owner. But the text of the plan, in Part IB of the format, can do no more than provide short summaries of the present state and treatment of the growing stock from information available elsewhere, e.g. in compartment histories and records of inventories, and a concise, initial analysis of the successes and failures of management to achieve current objects.

But the short summaries of facts and analysis of the current management position must suffice to show the stage of development that the forest has reached and, in the light of the previous ecological assessment, its future potential and the broad lines which silvicultural treatment can follow to fulfil the owner's purpose.

This subdivision of Part I, therefore, is an essential preliminary for use later in the first chapter of Part II where the foundations for future policy and action are laid. But Part II should not be anticipated. Here the opinions and conclusions depend on the silvicultural state of the forest relative to current objects of management, treatment and ecological potential, but exclude a study of wider influences such as those of market trends and resources of finance, staff and manpower.

Silvicultural assessment

Past treatment. Summaries of past treatment should aim to show the trend and phases of silvicultural development that the forest has experienced and thereby explain its present state. The account should therefore summarize treatment during periods of forest neglect or destruction, main programmes of reservation and enclosure and main methods of silvicultural working.

The forest types. The description of past treatment should lead naturally to the understanding and description of the present state and distribution of the growing stock which is best classified by forest types according to primary structure and then by species, thus:

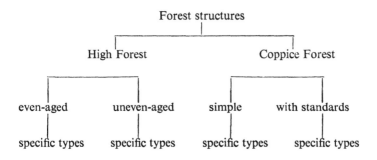

The degree of subdivisions into specific forest types will depend on their individual peculiarity and importance and the intensity of management. The extent and general distribution of the forest types are then described with emphasis on their present health, vigour, silvicultural prospects and the main methods of regeneration and tending used. The object is to summarize the variety and condition of the growing stock as it is now from a silvicultural aspect.

Quantitative assessment

Stock-mapping. The stock-mapping stratifies the forest by selection and delineation of the stand types which depend on the initial classification of forest types sub-divided into age, treatment or other classes. Thus even-aged Douglas fir might be separated into stands by five-year age classes which together form the Douglas fir even-aged high forest type. In the stock-mapping it may also be necessary to distinguish and delineate separately stands that have been wind-blown, heavily exploited or disease ridden while in intensive forestry sub-classification by yield or quality class may be necessary. The stock-mapping thus provides a stratification for measurement of areas, volumes and increment and is the basis of an inventory.

Inventory results. Inventory results should be summarized for units as large as is possible consistent with uniformity of the forest or silvicultural management types. Thus figures for irregular, selection forest must clearly be kept separate from figures for even-aged forest. Similarly in even-aged forest figures for units with markedly different rotations must be kept distinct as it is clearly no use to amalgamate the age-class figures of oak with a rotation of 150 years with those of Douglas fir with a rotation of sixty years; to do so would give an entirely erroneous impression of the state of the forest unless treatment instead of age-classes were to be used. There would, however, be little harm in amalgamating figures for two or more species whose average rotations varied by only 10 or 15 per cent. Therefore summaries must be for distinct forest types or for separate working circles (or even separate species within a working circle).

If a forest (or working circle) has been under one form of treatment for some time and several comparable inventories have been made, graphs can usefully be made to show the changes that have occurred. An example of this has already been shown in chapter VIII for both volume and increment by main size-groups for the selection forest of Couvet. By substituting standing volumes and volume increments of age-classes or age-groups for the size-groups used at Couvet similar graphs could show the progress of an even-aged forest with an additional graph to show the areas of each age-group or class.

A convenient form in which to tabulate inventory results by age-classes for an even-aged forest or felling series is shown in Table 24.

Table 24

INVENTORY RESULTS BY AGE CLASSES

Age-class	Volume per acre h. ft.	Increment %	Increment per acre h. ft.	Area acres	Volume h. ft.	Increment h. ft.	Normal area acres
TOTALS							

If the figures can be estimated columns could be added to show the normal volumes and increments for a stocking comparable to that achievable in practice.

Summaries of past yields, if possible by the same territorial units, make a valuable addition to the summaries of the inventory. If they are not available for the same units, their amalgamated total for the whole forest and by its administrative sub-units for comparison with similarly consolidated figures of inventory results still have substantial value. It is always important to know the relation between increment and the actual yield cut.

Methods of mensuration

Although methods of measurement used to compile inventories will not be elaborated here some comment is necessary since the manager should always define what information and what degree of precision he wants so that the compiler of the inventory can decide the methods of mensuration to be used and organize the work to obtain the precision and detail required in each forest and stand type.

As a general principle, in accordance with the intensity of management, the detail and precision wanted will vary directly with the value of the stands or individual trees and their nearness to maturity. Additionally, the need to know the contents and increment of individual territorial units (e.g. the block or section, compartment or sub-compartment) also increases with their value and maturity. The degree to which volumes should be classified by size assortments will depend largely on the types of market demands.

Accordingly in even-aged forests, even for intense management which uses a small sub-compartment as the unit of working, it is seldom necessary to make detailed measurements of stands below half or three fifths of rotation age; the younger stands can be assessed by a low intensity of sampling and application of yield tables. But for intensive management not only is a higher intensity of sampling needed in stands above half rotation age (to obtain a precision within, say 5 per cent and even less for mature stands), but the method used should ensure the sampling of each working unit such as the sub-compartment to provide figures for it. With less intensive management or for very uniform site conditions it may suffice to sample only the stand type and apply the type figures to the various working units.

In uneven-aged forests the same principles apply but the lack of area as a factor to indicate the distribution and quantities of maturing timber and the lack of yield tables affect the conclusions. There is an essential need to enumerate and measure small trees to obtain adequate figures of the rates of recruitment from below. Consequently the lowest measurable size in intensively managed selection forest in Switzerland is twenty or even fifteen centimetres diameter at breast height (equivalent to about $\frac{1}{4}$ or $\frac{1}{5}$ of exploitable age). Similarly the intensity of sampling will be high or even total as is usual in Switzerland. In extensively managed tropical forest the lowest measurable size should be at least one or two classes and preferably three or four classes below the minimum exploitable size, although it may be practicable to measure only the more valuable species.

To obtain a given precision the intensity of sampling will depend on the uniformity of the population and the area sampled—a greater sampling intensity is needed for both smaller areas and less uniformity. The manager should therefore decide, before measurement begins, for each forest and stand type what precision is acceptable and what is the range of areas of the forest working units for which separate estimates are wanted. To the precision needed the details of information wanted must be added—increment, volumes and their assortment by sizes (and quality if necessary), range of diameter and so on.

Management assessment

Past working plans. Accounts of past working plans, very short for old ones but more expansive for recent ones, particularly the one now under revision, should summarize previous objects of management and main methods of working to show the trend of silvicultural management and illustrate major changes in policy and method to attain the objects. It should not overlap the earlier description of treatment which is designed to explain the present distribution, type and health of the growing stock.

Analysis of silvicultural management. This analysis should show how the state of the present growing stock and its treatment fulfil the current objects of management and should emphasize the matters that need correction. Such an appreciation should be constructive by showing what alternative methods might be used such as change of species, thinning cycles or intensities, realization of overmature stands or correction of the age or size-class distribution, and change of rotation or removal ages. A comparison of the rate of cut with the increment in relation to the volume of the growing stock desired is a valuable aid to this analysis.

Although it is an appreciation of the silvicultural management only, economic or other factors may be noted to be the reasons for arrears in, for example, the thinning programme or early tending, accumulation or lack of mature stock and the like.

C. The administration and economy

A working plan cannot be expected to give more than a summary of the administration and utilization of forest products and of the economy by picking out the salient and critical features and analysing constructively the efficiency of the present management and the trends of markets and prices. Full information and analysis of prices, costs, use of equipment and so on should be collected and recorded as an integral and routine part of ordinary management and be kept under constant review for constant use in adapting practice. The working plan merely gives an opportunity for a periodic summary of the state of affairs, a summary which is valuable not only for comparison with past summaries as a guide to the trend of events but also for stimulating special thought on the need for changes in policy and practice.

A forest enterprise as a commercial business

This subdivision C of Part I of the plan is, in effect, a business summary of the enterprise and postulates that a forest enterprise should be run as an efficient business concern whether it is being managed entirely or mainly for profit or not, whether it is owned by the State, a company, private individual or any other body. The technical aspects of forest management, i.e. the silvicultural organization and manipulation of the growing stock which forms the main subject matter of this book, must be bound to the business of running the enterprise without which the application of the technical planning will be inefficient and by its inefficiency provide for its own disuse and destruction. The continuance of forestry practice depends on the success attained in achieving its objects and that success, however brilliant the technical forestry planning may be, depends ultimately on efficient operations which must, therefore, be linked with sound business principles and methods. But conversely, however brilliant the business operations may be the enterprise will also fail if the silvicultural side of the concern is poorly handled. The business management must combine efficiently the several branches of a forest enterprise—silviculture, logging, road and other engineering, sales and the rest—into a coherent efficient whole to achieve the owner's purpose which in its turn depends ultimately on the consumer's wants. (If the owner, for example, wishes to manage his forest solely for his own recreation and pleasure, the consumer of the products of the forest is the owner himself.)

In an appreciation of a forest enterprise as a business there are two basic aspects to consider, namely, the wants of consumers for forest products (services or material goods) and the resources of the forest owner to satisfy the wants. The resources (composed of land, labour and capital) comprise the 'input' whereby the yield or 'output' is produced at a cost to the owner to satisfy the wants of consumers at the prices they will pay. The manager of a forest enterprise, therefore, must always endeavour to keep the cost of the input as low as possible. By doing so it may be possible either to lower the acceptable price for each unit of output thereby benefiting the consumer who is then likely to increase his demands, or to maintain prices and thereby increase the favourable gap between costs and prices for the direct benefit of the forest owner. Even without elaborating the economic influence of costs and prices on the relations of supply and demand it is clear that the control of costs of input is vital to the proper management of a forest.

Subsequent discussion will, therefore, be related to the two basic aspects of wants and resources.

Wants. The object here is to give information which will enable the forest owner to decide what goods or services should be produced to serve his aims, such as highest profit, services to the community and so on. The types and quantities of forest produce wanted, therefore, should be summarized together with the prices that the market will pay analysed in relation to what the forest can produce (as shown in Part IA of the plan). It is very useful to establish any gradient of prices relative to size or quality of produce. The analysis should also consider the trends of demand and prices in relation to the trends of future supplies from the forest and from other competing forests.

A subsequent comparison (in Part II of the plan) of prices with costs of producing the products (analysed under resources) will enable the owner to decide what to grow in future.

In summarizing demands and markets it should not be forgotten that in many circumstances the supply of a material demand, for example firewood in a remote but populated, underdeveloped region, may be a matter of essential welfare for the consumer whether the prices paid are economically acceptable to the forest owner or not. This feature of demand should be stressed as its importance may be paramount. But the cost of satisfying the demand and possibly the pecuniary loss suffered by the forest enterprise, should also be shown; the difference between the economic price and the actual price paid is then, in effect, the cost of supplying a service although in fact material goods have been delivered—at a favourable and uneconomic price.

Resources. The object is to analyse the resources or means of production to enable the manager to increase efficiency and reduce the cost of all products and to help him distinguish the more profitable types of product. Much therefore depends on a review of administration and main methods, of cost analyses and of method and work studies to show where and how accepted methods and organization of staff, manpower and equipment can be bettered. It is seldom that the working plan officer will have time and opportunity to collect such figures of costs or method studies himself; as stated earlier he will have to rely on studies already made in that forest enterprise itself or by research elsewhere. Nevertheless, by his general study of the organization, methods and costs as they are, the working plan officer should be able to distinguish the fields where re-organization of staff and manpower, change of methods or modification of equipment seem necessary. His study may also show and emphasize a lack of pertinent information, owing to neglect of cost analyses or inadequate records or accounting procedure, and the scope for improvement.

Resources can be conveniently summarized and discussed under three main heads—administration, the growing of forest products and the utilization or harvesting of the products. The growing of forest products has already been discussed in Part IB except for the costs of the several operations or jobs entailed for which space has been provided in the format in chapter IV of Part IC.

It might be reasoned that ideally records should be kept to show the cost of a unit of each product delivered to a buyer or other consumer. Only then can a fair comparison of costs and prices be made. In many enterprises the cost of a product can be calculated from cost analyses which show the cost of each manufacturing process to which is added the appropriate share of overhead costs. But for forest products there is almost always a long lapse of time between initial costs of formation and harvest. Interest must be added to those costs. But a more serious difficulty is that it is not possible to apply *present* costs of formation and tending to a stand planted years ago—methods, costs of labour and the value of money may all differ greatly from those experienced perhaps fifty or more years ago. Consequently it is necessary to have a record of actual costs and the time when they were incurred, stand by stand. Even then the application of annual administrative, protective and other costs during the life of a stand is fraught with difficulties. Some average figure might be applied, such as the mean of present overheads discounted to the date of formation and over-heads current at that date. But in practice results, even if an acceptable precision is obtained, would not justify the expense of keeping the detailed cost figures necessary for each and every stand or other working unit. Possibly, as a matter of economic research, they could be kept for sample site-types and species. But in any event the past is over and cannot be changed. Decisions on when to fell a stand do not depend on a comparison between the cost that has been incurred to grow it and the price now obtainable for it but rather whether it is more profitable to leave it to grow longer or cut it now. It is present and expected prices that are the crux of the matter.

Nevertheless a record of past prices and sample calculations of the costs that have been experienced in growing stands and the consequent profit earned by them demonstrates the success or failure of past management to make a profit and also serves to show the trend of prices.

Financial statements. The financial statements should include the current profit and loss accounts and a balance sheet showing the present state of all assets and claims on the forest enterprises with critical comments on them. These matters are discussed later in this chapter on page 294 *et sqq.*

PART II. THE FOUNDATION OF THE PLAN

This part of the working plan constitutes the heart of the planning on which all else depends. It must, therefore, contain both an appreciation of the current position in the light of future prospects to provide a base on which to confirm or revise the objects and policy of the owner, and a study of alternative courses of action culminating in a choice of the one or more courses that are deemed to be the best. The decisions made here are far-reaching and must be truly effective and long lasting; their subsequent change in later years should be made only with reluctance after full justification. Yet the decisions made may be controversial or liable to misunderstanding now or later if they are not justified by cogent reasoning supported by a clear exposition of facts and forecasts and by analyses of alternative policies and methods. Therefore some degree of expansiveness in the reasoning, without lack of lucidity and conciseness, is desirable although details of supporting figures and calculations may be allotted to appendices.

Nevertheless in exposing the present position and prospects in chapter I there should not be an undue repetition of the data and reasoning already given in Part I of the plan but rather a consolidation of the salient and critical features in a reasoned summary with references to Part I for detail. Exceptionally in Great Britain, where only this part of the plan is seen by the headquarters of the Forestry Commission and where, after being sanctioned by that high authority, this Part II serves as the directing instruction for the consequent action and short-term planning controlled and sanctioned by lower authority, greater length and more detail may be necessary to make this Part II a coherent whole complete in itself.

The essence of the long-term policy is that it should last. Therefore decisions made must be sound, far-sighted and clearly enunciated without confusion of what is an object or purpose with what is a means of accomplishment. Thus a true object may be the attainment of the highest financial yield. To accomplish that object it may be decided to grow coniferous saw-timber, perhaps of a definite species to a particular size or age. But to grow saw-timber is then a means to an end, not the object; in other conditions the same object could be attained by growing coniferous pulpwood or chestnut coppice.

Therefore an object may be constant throughout a working plan area but the means of its fulfilment might vary according to site conditions. On the other hand the objects themselves may vary with type or position of site—for example, the supply of essential local agricultural needs in the lowland fringes, recreation and amenity in an accessible, watered valley and the highest financial yield elsewhere. The zone of application of each object of management is an essential part of policy.

The long-term policy should, therefore, define the objects, the zone or type of land to which they apply, and the broad types of product considered to be the best means of accomplishing each object. If multiple objects apply in one area, such as highest financial yield and sustained yield, or amenity and highest possible net income, the degree of priority to be accorded to each should be made clear.

Conclusions reached must be justified and also conform with constraints and commitments. Thus the growing of coniferous saw-timber to achieve the highest financial yield must be shown by calculation to be better than growing other produce such as pulpwood; the policy adopted in a State forest must be shown to agree with general government forest policy; commercial exploitation of timber must be shown to conform with the supply of right-holder's needs.

The long-term policy should also outline the types of forest wanted and their silvicultural organization, the intensity of working, development of communications and administrative arrangement. To do so will entail decisions on the silvicultural systems considered to be the best for the several objects of management and the phases of conversion to those systems and their ideal states that are judged necessary together with recognition of the size of the unit of working, such as the compartment of sub-compartment and its minimum size in order to identify the intensity of management desired. Linked with the working unit and intensity of working is the policy of road intensity desired and consequent programmes of road construction. The intensity of working also affects the administrative policy and the distribution of staff and size of their charges. Without direction on these matters that define the ideal forest enterprise aimed at, the phases of conversion and the intensity of working that are foreseen, confusion in settling the consequent short-term plan will arise. But the long-term directions need only outline these matters in broad but inequivocal terms.

The framework of the plan

Having determined the objects of management and the basic methods for their fulfilment it is necessary to interpret and summarize the action implied by defining the framework of the plan in concrete terms of territorial units to show their distribution, position and extent. It will probably be necessary in large or heterogeneous forests to have two or even several working circles identified with objects of management and silvicultural system or main forest types all of which may depend on site conditions of position relative to markets, or of topography and soil types. It may also be that some working circles should be subdivided into felling series which may coincide with sub-managerial charges.

It may often be possible to summarize matters in tabular form without much previous explanation since the principles on which division of the area depends should have been enunciated already. One table may suffice to relate the area and position of working circles and their felling series with administrative units and forest blocks or enclosures and even compartments but the inclusion of sub-compartments here becomes unnecessarily detailed. An example with hypothetical entries is given in Table 25 to show how the whole of a working plan area should be accounted for, without reference to compartments which can be given in the main area statement in an appendix (see example in Table 30).

Table 25

SUBDIVISION OF THE WORKING PLAN AREA: ALLOTMENT TO WORKING CIRCLES AND FELLING SERIES

(areas in acres)

Administrative charge	Forest Block or Enclosure	Area of working circles by felling series						Unallotted areas	Total area	Remarks
		Hardwood W.C.			Conifer W.C.					
		F.S.1	F.S.2	Total	F.S.1	F.S.2	Total			
I	A	510	—	510	—	205	205	15*	730	*Nursery
	B	—	—	—	—	870	870		870	
	C	705	—	705	70	155	255		930	
	D	270	—	270	50	640	690	10*	970	*Depot and Offices
	TOTAL	1485	—	1485	120	1870	1990	25	3500	
II	A	—	240	240	445	—	445		685	
	B	—	310	310	405	—	405	25*	740	*Leased for Agriculture
	C	—	170	195	520	—	520		715	
	D	60	160	220	260	—	260	30*	510	*Unproductive marsh
	E	40	140	180	315	—	315		495	
	TOTAL	125	1020	1145	1945	—	1945	55	3145	
GRAND TOTALS		1610	1020	2630	2065	1870	3935	80	6645	

Notes (i) In both Hardwood and Conifer working circles parts of felling series No. 1 occur in both administrative, sub-managerial charges for reason of convenience of extraction to markets.

(ii) It is too expensive to drain the marsh to make it productive but it has botanical and other scientific interest.

(iii) The areas of roads and rides and water-ways have been included in the production area of working circles. No waterway is more than 15 feet wide, no road or ride over 30 feet including ditches.

PART III. MANAGEMENT OF THE GROWING STOCK
IN THE PERIOD OF THE PLAN

The next part of the working plan applies the long-term policy and the broad definitions of main methods and phases of progress to action in the short-term namely during the period of the plan. This part of the plan, therefore must be more specific and eminently practical but less explanatory than Part II since the main reasoning and justification of basic decisions have already been made there. Nevertheless it must contain enough explanation to prevent doubt and ambiguity in execution. To be more specific and to enhance clarity and precision of expression it is necessary to distinguish between treatment of the growing stock (in Part III) from other management of the estate as a whole (in Part IV) and to treat each working circle separately in distinct sections A, B, C, etc.

Some amplification, adaptation or even repetition of the precepts of Part II may, however, be necessary in order to apply them clearly to the particular conditions of each working circle and to include the detail required for actual application in the field. Thus in the *management design* a detailed tabular statement, in extension of the one used in Part II, is necessary (even if confined to an appendix) to show the exact limits of the working circle, down to sub-compartments if they are used. A management map in an appendix will also support the tabular statement. Also the particular objects of management that apply and their influence on the type of products to be grown require clear definition with an explanation of the priorities and the relations between several objects. This definition and clarification of purpose should enable the silvicultural plan for the working circle to be designed—e.g. the choice of species and silvicultural system, adjustment of age classes, thinning intensities and so on.

The *silvicultural design* will almost certainly require one or more tabular statements or graphs and some introductory description of the constitution and structure of the growing stock since the summaries in Part I of the plan may not be specific to the working circle now concerned. In any event, the peculiarities of the growing stock with reference to the objects of management require analysis—e.g. maldistribution of age or size-classes, low increment or potential of some species or their unsuitability to sites and so on. In even-aged forestry it is necessary to elaborate the principles on which is based the selection of stands to be finally felled and regenerated (the regeneration block or area), principles which must be connected to the objects of management, rotations and standards of maturity (removal ages) and the adjustment of the ratios of age-classes.

Similarly, explanation of the allotment of areas to establishment and thinning treatment is necessary so that the silvicultural design can culminate in a statement that shows areas allotted to the main types of treatment—final fellings, establishment, thinnings and possibly treatment preparatory to final fellings.

Lastly, the silvicultural design must show clearly what the unit of working is so that the intensity of treatment is defined, e.g. sub-compartments normally two acres or more in extent.[1]

For uneven-aged forests it is the exploitable sizes and felling cycle that need elaboration together with the intensity and pattern of tending fellings or cleanings in the smaller size-classes. Decision on the mode of tending to be adopted is particularly difficult and important in irregular, tropical forests where the multiplicity of comparatively useless species makes this work monetarily unrewarding yet potentially valuable and even essential both for future income and for shaping the growing stock structure and composition to a state nearer that desired. For example a long felling cycle and a small minimum exploitable size succeeded by intensive tending may be the first phase on the road to even-aged working.

Regulation of the yield

(For methods of regulation see chapter VIII.) Provided that the management and silvicultural organization and practices are happily designed and in harmony with methods of utilization and exploitation, regulation of the yield should be straightforward.

But the importance of yield regulation should never be underestimated. However flexible the prescription of the plan may be,

---

[1] This unit of working and intensity might be varied in size in accordance with treatment. Areas undergoing a period of rest or preparation for intensive treatment at regeneration may not need any classification by sub-compartments except for stock-taking.

however much discretion is given to the manager to decide from year to year what should be cut, it remains true that the yield in kind, quantity, quality, distribution and timing, is the basic factor for providing present income and for shaping the future growing stock, future income and the whole trend of the supply of goods and services. Therefore the greater the flexibility and discretion accorded to the owner the more necessary it is to define clearly the principles by which the various types of felling, in quantity, intensity distribution and timing, are decided. Even with very flexible working it is usually necessary to calculate the desirable quantitative yield, allotted as is appropriate to final and intermediate yields, with upper and lower limits which should not be exceeded. It is also usually desirable to allot areas or even stands from which the final fellings are to be selected, preferably in a recommended order of priority or sequence as a felling plan. With rigid working not only would the permissible range in quantity be narrow but the actual order in which stands would be felled might be prescribed, even to the year of felling—as, for example, in a coppice felling series—and incorporated in a definite felling plan.

Forecasts of outturn

The yield prescribed should not be confused with forecasts of outturn. Yields may be prescribed by area and give no indication of the quantities and sizes of wood that will be cut. If, however, the yield is rigidly prescribed by volume the actual amount cut each year should tally closely with that prescribed—at least in the early years of the plan period. But conditions are bound to vary from year to year owing to the effects of markets, wind or other damage, weather and so on so that the actual cut differs from the prescribed yield. Moreover to be fully useful forecasts of outturn should classify the volumes by assorted sizes and probably by species. Then the forecasts will facilitate not only budget estimates but work programmes and industrial planning which will be particularly important in regions of heavy afforestation or other forest development.

Forecasts of outturn, in addition to a prescribed yield, are therefore essential and depend on the yield prescribed and on conditions as they happen. They will consequently be reliable and precise only for a short period. Nevertheless long-term forecasts with less reliability and precision are also useful particularly in regions of forest development where the introduction of new industry or enlargement of existing industry will be entirely dependent on the prospects for their raw material.

In Great Britain where large scale afforestation continues after steady progress since 1919, the Forestry Commission lays great importance on these forecasts; assessment of outturn or production forms a large part of ordinary routine management as well as working plan procedure (Anon. Working Plan Code, 1965). The working plan officer prepares both short and long-term forecasts of outturn. The short term is for the next five years for each year of which the volume of timber proposed for cutting is classified by sizes for each of twelve species or groups of species, namely volumes to 3, 7 and 9 inches top diameter over bark. Yields from thinnings and final fellings are kept separate. This forecast is revised by the district officer or manager each succeeding year and extended by one more year. The figures for the first year of the forecast are as precise as possible and are supported by a detailed programme of work showing the exact position and extent of all operations for that year, by compartments or sub-compartments.

The long-term forecast, similarly classified, is for five, ten and fifteen and twenty years ahead but is revised quinquennially. To standardize procedure and obtain co-ordinated quinquennial forecasts for the whole country the revisions are made in the calendar years whose digits end with a 4 or 9.

The procedure used by the British Forestry Commission is designed for even-aged forestry and demands intensive management based on the detailed assessment and records of growing stock in every working unit which is normally the sub-compartment which may be as small as one acre. The information collected for each sub-compartment includes for each species, the age, area, volume (by diameter classes and per acre), top-height, mean girth at breast height and yield class. From the records each species is analysed for the whole working plan area to give its total area, area of each ten-year age-class sub-divided according to yield class so that a weighted value of the yield class can be calculated for each age-class and the whole area of the species. The procedure relies on comprehensive yield and production tables on which to base estimates of growth and yield after establishing the yield class of each stand or sub-compartment.

Although this is not the place to enlarge on British Forestry Commission procedure copies of some of their current forms, extracted from their working Plan Code, are placed in Appendix I to illustrate this brief summary of their forecasting procedure. The forms appended are:

WP 5 Area statement of growing stock
WP 6 Area analysis by yield class. A worked example
WP 17 Forecast of allowable cut
WP 18 Production programme and control
WP 20 Forecast of allowable cut. A worked example.

Silvicultural regulations

The felling and thinning plans which incorporate the quantities, positions and timing of cutting need to be supplemented with qualitative advice on how to cut. For clear felling it is obvious that no advice is necessary. But for regeneration fellings, for thinnings and all types of selection fellings guidance on intensity, species or stems and type of stand structure to favour is necessary.

Frequently silvicultural methods and techniques have become standard, local practice and are controlled by standard regulations. Then only reference to the regulations is necessary. But in their absence the working plan must include directions to guide the manager on how the fellings and thinnings should be made and how ground is to be prepared for planting or natural regeneration and so on.

Control of prescriptions

It is obvious that comprehensive records, based on well maintained compartment and sub-compartment registers or histories, must be kept to show what yields a forest has produced and where they were cut. But it is also necessary to compare annually the actual cut with what it was planned to cut. Not only does such a control show the extent to which prescriptions are exceeded or neglected but it also helps subsequent judgment whether the prescriptions were well devised for actual practice.

The form of control adopted will naturally vary but in many

circumstances a control of quantitative prescriptions may take a form similar to that in Table 26 for a final volume yield.

If sanction from higher authority is necessary to condone deviations from prescriptions, reference should be made in the remarks column to relevant correspondence.

Similar control forms can be devised for all other quantitative prescriptions, for example for thinning (perhaps by area), afforestation or road construction. But qualitative prescriptions are not amenable to control by recorded figures beyond a statement whether the work has been done or not but require field inspection.

The control procedure adopted by the British Forestry Commission is based on the forecasts. The forecasts implement the prescriptions made in Part II of the plan, but the quantities involved are constantly reviewed and revised according to

Table 26

CONTROL OF FINAL YIELD

cubic feet

Year	Volume yield pre- scribed	Actual volume cut	Difference		Difference to date		Remarks
			Excess cut	Deficit cut	Excess cut	Deficit cut	
1967	750	672	—	78	—	78	Slack markets
1968	750	685	—	65	—	143	Bad weather
1969	750	806	56	—	—	87	Good demand and weather
1970	750	835	85	—	—	2	Heavy wind damage
1971	750	775	25	—	23	—	—

circumstances. But the precise forecast made yearly by the district officer for the next year becomes the definite programme and prescription of work for that year and is approved as such by the conservator. Control, therefore, becomes a comparison of that programme with actual performance and not with wider prescriptions made at the start of the working plan period. There should therefore be little or no discrepancy as the programme, although complying with a prescription which may be qualitative or possessing a range of quantities, is made and approved in the light of circumstance each year for only one year ahead.

PART IV. GENERAL MANAGEMENT OF THE ESTATE

This part of the plan embraces a very wide field of essential activities and their subsidiary plans without which the most skilful management of the growing stock will fail or be inefficient and uneconomic. The implications involved in some of the activities described by the headings of the format (which is not exhaustive) maybe so great—e.g. nursery management and techniques or supply and control of grazing—that a complete chapter or even more would be needed to summarize their application. Moreover their individual importance and the degree to which they are treated in a working plan differs very greatly with conditions. Consequently only general guiding comments on some of the activities are made here.

Many of the activities may together constitute a major portion of the manager's daily work and will, therefore, be under constant review and often subject to standard practice and regulations. For example, many forests will have a fire protection scheme which through constant use is progressively tested and improved. Similarly nursery practice and techniques although affected greatly by research results and ordinary experience will seldom need revision in a working plan but only a statement showing the quantities and species of plants needed for the growing stock management and reference to alternative sources of supply. For such activities, therefore, the plan needs only to refer to the standard regulations and practices with recommendations for modification and change deemed necessary particularly those requiring new equipment or buildings, with broad estimates of costs and income involved.

Some activities, however, may require special attention in order to co-ordinate them with changes prescribed in the management of the growing stock or changes in policy and method enunciated in Part II. The implication of those changes on the several activities of estate management must be fully exposed. An obvious example is the construction of new roads or the reconstruction and special repair of existing roads and the plan should incorporate a definite programme for new roads, special repairs or reconstruction, planned in relation with the felling plan with a general estimate of costs.

Other activities may have become standard practice but require change to meet changes in policy or development of the management of the growing stock. Examples are methods of sale and exploitation, staff and manpower or mechanization. Broad justification of major change in practice should have been made in Part II and need not be repeated here; nor can details of big changes, such as from sales of standing trees to departmental felling, conversion and extraction and sale at central depots, be expected to be given in the text of the working plan. But an outline of the new scheme, completed in full and available separately, should show the changes required in manpower and equipment, expenditure and income as well as a programme or timetable for the process of change. Other less elaborate changes, such as those of staff, can easily be given in full so that old and new charges and costs can be shown in a comparative statement.

Some of the estate activities may be so extensive that they constitute the main object and business of management and consequently control or severely restrain the management of the growing stock. In such an event much more space is needed than that allotted in Part IV of the format of a working plan. For example, a forest estate may be heavily burdened with rights and concessions for grazing, fuel-wood and small timber. It may be that such rights are so heavy and so important that special working circles have to be formed to satisfy the right-holders' demands. For such a purpose a grazing working circle might be formed to produce grass, under controlled, rotational grazing and hay cutting, and also leaf fodder obtained from planted leaf-fodder tree species which are subjected to a lopping programme. Alternatively grazing may have to be allowed in other working circles managed primarily for wood production but subject to the constraint of grazing which might be excluded only from regeneration areas but limited in intensity elsewhere and favoured by growing a proportion of fodder tree species. Therefore, if rights and concessions are heavy special provisions may have to be made in Part III, but in Part IV provisions are made only for matters which do not have an appreciable effect on the objects and methods of treating the growing stock. Nevertheless it should always be shown in Part IV how the rights and concessions recorded in Part I are being satisfied by reference to the relevant prescriptions in Part III.

Records of the growing stock

A standard procedure which can become a matter of routine is desirable for recording the progress of the growing stock, the stock-in-trade of the forester. No business can be successful without a full knowledge of its stock at all times and the working plan of a forest should ensure that the means of recording changes in the growing stock are adequate.

The author of a working plan, especially the first plan for a private owner or commercial body, may have to design the forms to be used for record keeping. A working plan for a State forest however, will usually have to conform with standard departmental practice that has been decided already to suit the several kinds of State forest and to agree with accepted accountancy and other administrative procedure. But that in no way implies that the record keeping of the growing stock is a subsidiary matter. On the contrary it permeates many aspects of forest work and planning and its importance is such that other procedures such as accountancy ought, perhaps, to be adapted to methods of recording the growing stock rather than *vice versa*. Ill-designed records of the growing stock or constantly changing designs will hamper analysis of the success or failure of treatment, assessment of growth and future yields, and decisions on future treatment and even assessment of the true financial state of the property.

The advent of calculating machines and electronic computers has made it comparatively easy to use records so that the demand for them constantly increases. But that truth must not obscure the danger of excessive collection of facts since their collection in the field and entry into forms in the office is time-consuming and possibly demanding of skill. It is always useless to prescribe any action which is beyond the capacity of the staff to perform. The information to be collected, therefore, should be confined to that which is really necessary for the purpose, methods and intensity of management and the forms should be as simple as possible. Nevertheless the prospect that developments in forest management, in intensity or scope, may require more detailed records than now wanted should not be overlooked. Provided that the work is within the capacity of the staff additional details are likely always to prove their worth.

Compartment histories or registers

Being the smallest permanent territorial unit in a forest, the compartment becomes the unit of permanent record. But, if the unit of working is the sub-compartment, temporary though it

10

may be, it is preferable for it to be the basic unit for collection of facts, which are then amalgamated for the containing compartment or other unit. That procedure is particularly important if, owing to variety of site and forest types, different parts of a compartment are allotted to different working circles and treatment. Sub-compartments however express changes in site values by differences of species and yield class and would always be wholly within one working circle. After prolonged management treatment may become stabilized so that compartments could be re-organized so that each is wholly within one working circle. Even then variation of species, age or yield class within a compartment will make sub-compartments necessary and provide distinct populations that require separate sampling and assessment.

It may therefore be concluded that the system of record-keeping must be based on the sub-compartment as the unit for collection of data which can then be allotted to the containing compartment and working circle and can be related to the treatment needed.

Type of information

The purpose for which information is collected is basically threefold, namely to record what is there, to judge the effect of past treatment and to assess growth and yields for future treatment. Therefore the type of information wanted must vary with the method of treatment which depends on the basic silvicultural system used. If the system is uneven-aged the element of mensuration and record is the individual tree identified by its size; if the system is even-aged the element is the stand identified by its age and area.

Therefore for an uneven-aged system it is necessary to classify the frequency of stems in selected size-classes by diameter or girth, separately by species where necessary. The volumes of each size-class can also be recorded. Then, if the method of control is used, the repeated comparable inventories of both standing and felled trees enable increment (and times of passage) to be calculated. Results can be recorded compartment by compartment and thence for the felling series, working circle or forest as a whole in the manner shown for the forest of Couvet in chapter VIII.

If the method of control is not used it will be necessary to measure increment of the size-classes by borings or repeated measurement of sample trees and, preferably, also record volumes of casualties (fellings, deaths, etc.). Records similar to those of the method of control can then be established.

For even-aged systems the information needed will be for each stand. The basic information needed (for which the record forms must provide spaces) includes: area, species, age, density of stocking, mean breast height size, mean height (probably top-height), yield class and preferably volume and increment also. With that information full use can be made of yield and production tables if they are available. If they are not available volume and increment must be measured and recorded.

Obviously information collected and recorded must be consistent for the several stages of their use from the sub-compartment or stand to the whole working plan area and must agree with methods of working and the yield, volume and other tables in use. To illustrate the necessary co-ordination examples of the sub-compartment and compartment forms (WP. 0 and 1) now used by the British Forestry Commission are appended in Appendix I.

It will be seen that volumes are not included in those two forms. That seems a pity but it is perhaps unnecessary owing to the importance given to area, age and to the general yield class, modified if necessary to the local yield class, which is a basic factor in a management which depends greatly on the normal yield production and thinning control tables recently completed for the sixteen species commonly used in Britain. It will also be seen how the data recorded in the sub-compartment and compartment forms are extended and applied to complete the forms WP. 5, 6, 17, 18 and 20 which are also shown in Appendix I.

A full account of the tables and their use is given by Bradley *et alii* (1966) in Forest Management Tables and in instructions in the use of the forms in the Forestry Commission's Working Plans Code (Anon. 1965). Here only a very short summary in the form of notes is given, adequate it is hoped to explain broadly the forms, tables and methods of management used:

1. In general the State forests of Britain are managed to obtain the best economic results (but see also chapter II under British Forest Policy) for which it is found suitable to use rotations at which the MAI culminates after thinnings whose

intensity produces the greatest girth increment without reducing volume production.

2. The majority of British forests are the result of afforestation and reforestation since 1919 so that they are mostly in the thinning stages and final yields are as yet small.

3. It has been found that, during the normal thinning period, steady thinnings which remove 70 per cent of the maximum MAI satisfy the objects given in (1) above, irrespective of species, yield class or age (within the thinning period). That implies the removal of about 50 per cent of the CAI during the same period, i.e. during about $\frac{3}{5}$ of the rotation.

4. The yield classes used in the normal yield and other tables are based on the maximum mean annual volume increment whenever that may be reached. Thus both Scots pine and sitka spruce in the 160 yield class will have a maximum MAI of 160 hoppus feet/acre, but the former species attains it at an age of sixty-six and the latter at fifty-seven years. The thinning intensity of both would be 70 per cent of $160 = 112$ hoppus feet/acre/year or 336 hoppus feet/acre every third year on a three year thinning cycle.

5. Consequently thinnings are controlled by volume at 70 per cent of the MAI indicated by the yield class allotted to each stand, the allotment being based on the top-height of the stoutest forty trees per acre relative to age. Thinning control tables facilitate the check.

6. Therefore it is possible to forecast production, namely the thinning yield plus final yields which are the standing volumes at removal age, normally the rotation of maximum MAI. But the forecast of final yields from tables is not very reliable and is usually based on measurement. Production forecasts can be classified by size assortments.

7. The production forecast tables use the gross area of the sub-compartment or compartment, i.e. inclusive of rides, and the thinning control tables use the net area. The tables using gross area therefore must have lower figures per acre than those using the net area and it is assumed that the figures per gross area average 85 per cent of the net area. So the production forecast for thinning is 15 per cent less than the thinning control figures, i.e. 15 per cent less than the maximum MAI, to allow for rides, incomplete stocking and so on. In effect, therefore, for the gross area the intensity of thinning is $85 \times 70 = 60$ per

cent of the yield class or maximum MAI. This adjustment is necessary as the areas of compartments are recorded as their gross areas, inclusive of rides. But the control of thinnings, i.e. by sample plots in the stand during marking, must clearly use the net or exact area of the growing stock treated.

To enable the reader to follow the whole process from the inventory and records of the sub-compartment (Form WP. 0) to the area analysis of the work plan area by yield classes (WP. 6) or to the Forecast of allowable cut (WP. 20) and to see the types of tables used, the following British Forestry Commission tables are appended in Appendix I. (*Note*. The tariff numbers shown, e.g. in the thinning control table, refer to the tariff volume tables used.)

List of appended tables

(*a*) Thinning control tables—Sitka spruce, yield classes 100 to 180
(*b*) Production forecast tables:
    (1) Thinning yields—sitka spruce, yield classes 140 and 160
    (2) Felling yields—sitka spruce, yield classes 140 to 160
(*c*) Normal yield tables:
    (i) Sitka spruce, yield classes 160, 180, 200
    (ii) Douglas fir, yield classes 160, 180, 200

Table 27

COMPARTMENT RECORD OF QUANTITIES CUT

Date	Sub-Compt.	Species	Mean B.H.Q.G.	Volumes cut, hop. ft.		
				To 3 inches top over bark	To 7 inches top over bark	To 9 inches top over bark

Study of the forms will show that much depends on comparatively wide use of standard methods of working even-aged plantations and that the basic units used are area and yield class and, to a smaller extent, age. A noticeable omission in the compartment forms is provision for recording quantities cut.

Although such entries involve considerable clerical work and measurement in the field the lack of the information seems serious and may be an argument for having much larger compartments or parcels as proposed in chapter VI.

Possible headings for such a record of the cut are given in Table 27.

Sub-compartment details might be omitted if such sub-division is impractical in the field.

Finance

The finances of an undertaking are always an important and often a critical element. Therefore a working plan should summarize the present financial position and show what changes are expected to occur as a result of executing the plan.

The balance sheet. The financial state of an undertaking is summarized by the balance sheet which records the assets or values of the resources of the business and compares them with the claims against the business. The claims include firstly all liabilities of the business, i.e. obligations and debts to others, and secondly the net value of the business, i.e. the difference between assets and liabilities, which constitutes the obligation of the business to its owners. Total assets therefore equal total claims.

The variety of headings used in a balance sheet will naturally vary with the size and complexity of the business but a simple one is shown in Table 28 to show the types of headings used. Fixed and current assets are broadly distinguished by the time that elapses between acquisition of the asset and its discharge. Thus, as a rough approximation, if the period is less than one year the asset may be said to be current (Edey, 1966). The same distinction can be applied to liabilities. Consequently (as examples) machinery is classed as a fixed asset since cessation of its use, dismantling and sale is usually slow, and all stock such as work in progress or the finished article is classed as a current asset since it can be quickly sold.

The balance sheet in forestry. The preparation of a balance sheet for a forest enterprise requires decisions firstly on how to classify the growing stock among assets and secondly how to estimate its value. As pointed out by Gane (1966) the growing

Table 28

SAMPLE BALANCE SHEET OF FIRM 'X' — 1ST JANUARY

CLAIMS	£	£
1. *Long and medium term liabilities*		
6% Debentures	5,000	
Mortgages	3,000	8,000
2. *Current liabilities and provisions*		
Creditors	2,700	
Bank overdraft	1,550	
Taxation	6,950	
Provision for shareholder's dividends	3,250	14,450
		22,450
Shareholders' interests		
3. *Issued share capital*		
Preference shares	6,500	
Ordinary shares	24,400	30,900
4. *Capital reserves*		5,200
5. *Retained profits*		4,300
		40,400
SHAREHOLDERS' (OWNERS') FUNDS		£62,850

ASSETS	£	£
1. *Fixed assets*, at cost depreciation		
Lands, buildings and roads	25,600	
Plant, machinery, office equipment	7,100	
Motor vehicles	10,260	42,950
2. *Current assets*		
Stocks	10,500	
Raw materials (£2,500)		
Work in progress (£6,200)		
Finished goods (£1,800)		
Debtors	7,500	
Investments (market value)	1,150	
Bank and cash balances	750	19,900
TOTAL ASSETS		£62,850

stock of a forest is a hybrid between current and fixed assets—
it is partly stock (though standing and not ready for disposal,
and so perhaps classifiable as work in progress) and therefore
partly a current asset and partly the machinery of production
(as shown in chapter I) and therefore part of fixed assets. As
stated by Gane it is not merely a matter of terminology since
fixed assets do not appear in the profit and loss accounts but
stocks (current assets) do and are traditionally valued at cost.

It may perhaps be accepted that the asset of the growing
stock should be recorded separately from either fixed or current
assets provided that their value is assessed in a way which is
practical and indicative of changes in the growing stock,
changes which can then be incorporated in the profit and loss
account. It may be noted that the changes will usually reflect
appreciation (less removals) rather than depreciation. The
growing stock value would be recorded in the balance sheet
after fixed assets and before current assets.

Valuation of the growing stock. The practice of the Forestry
Commission (1966) in Great Britain has been to value the growing stock at the net cost that has accumulated in growing it.
Such a method implies that a plantation only ten years old
is valued at the cost of its formation *plus* administrative costs
for ten years *plus* accumulating interest—a total of perhaps
£100 the acre. Yet the market value of the trees may be nothing.
In the early years of a plantation the cost value per cubic foot
of wood will be very high, a value per unit which will steadily
fall with age as long as growth accumulates faster than the growth
of costs (as influenced by compound interest). In complete
contrast the commercial value of wood per cubic foot tends to
rise with age. Moreover there is the difficulty of choosing the
correct rate of interest for calculating costs.

Hiley (1954), recognizing the need to include in the profit
and loss account the appreciation in value of the growing
stock by annual growth, adopted a method that used increment
for the management of the Dartington woods in Devonshire.
His method for merchantable timber stands, applied to each
sub-compartment, was to increase yearly the assessed value of the
stand (at inventory) by a percentage which was subjectively
decided as a proportion of the real increment to allow for
removals by thinnings. It was considered to be too complicated

to include thinnings in the adjustment. But new plantations were valued at cost, values which were then increased by a subjectively decided percentage of not more than $3\frac{1}{2}$ per cent, so that their book value roughly equalled their market value when the plantations become merchantable.

In his study of the matter Gane (1966) also shows that increment must be used in assessing appreciation of value. He proposes that market prices should be applied to the increment with due allowance for difference in prices obtained for thinnings and final fellings. The proportion of increment removed in thinnings is valued at the rate obtained for thinnings of the size or age concerned but the value of the increment retained (i.e. remaining in the stand after thinning) is assessed at the value of the mature timber that it is destined to produce. The total value of the increment of a stand is thus calculable. For the calculations of the value of the increment stands are classified into suitable categories of species, yield and age class. By using yield tables (or local increment assessments) a practical procedure of calculating the value of the increment of a forest is possible. This incremental value is added to the initial assessed value (at inventory) of the growing stock and the value of all removals subtracted to give the yearly estimated value. In calculating the value of removals Gane proposes that the same, book values used for the increment should be used for the removals rather than actual market prices received each year.

Gane's proposals appear practical and not too difficult to apply after their first introduction particularly for a forest composed of only a very few species on uniform sites. But if the growing stock is variable, in species, site and methods of treatment his method may be more complicated than is really necessary if inventories are made at ten-year intervals or less. It would seem adequate to apply annual increment and exploitation values to the value of the growing stock as assessed at the inventory, as described below.

At each inventory the volumes of the growing stock would be measured and recorded—stand by stand, compartment by compartment or by other stratified unit suitable to the intensity and method of management. The net value of the standing volume of each unit (according to its species, size, quality and location) is also assessed and recorded. By addition we then have figures for the total standing volume (AV) and value (VG) of the growing stock and also by calculations the average, net standing value (v) of the timber per cubic foot. This average value per cubic foot allows for variation in species, size, accessibility and so on. At the same time the increment of each unit is assessed to give by sum the total increment of the forest. This annual increment is expressed as a percentage (i) of the total growing stock volume (AV). Each year the volume of removals (from thinnings and final fellings) is recorded and expressed as an exploitation percentage (e) of the growing stock (AV).

Then where AV_1, AV_2, etc., are the volumes and VG_1, VG_2, etc., are the values of the growing stock in successive years after the inventory

$$AV_1 = AV + \frac{i \cdot AV}{100} - \frac{e \cdot AV}{100}$$

and

$$VG_1 = v\left(AV + \frac{i \cdot AV}{100} - \frac{e \cdot AV}{100}\right) = vAV\left(1 + \frac{i}{100} - \frac{e}{100}\right)$$

$$= vAV\left(\frac{100 + i - e}{100}\right)$$

The yearly change in value of the growing stock then equals $VG_1 - VG$ and in successive years $VG_2 - VG_1$ and so on.

By assuming that the increment itself and the average value per cubic foot at the inventory continue unchanged between inventories and by assuming that value applies equally to the increment and exploitation, the annual calculation of changes in value is simple however varied the growing stock and conditions may be. Refinements could be introduced by calculating separate value rates for increment and exploitation but it seems doubtful whether the extra complication is necessary provided that intervals between inventories are not long.

Profit and loss account. To provide the figures of profit or loss for inclusion in the balance sheet a profit and loss account which summarizes the finances of the various operations of the enterprise must be kept whereby a classified analysis of all items of income and expenditure displays the net profit (or loss) for the year or other period. In addition the process of summarizing income and expenditure leads to a study and critical analysis of the various components and their relative influence on the profit or loss so that the best field for reducing costs and

Table 29

SAMPLE FOREST OPERATIONS AND PROFIT AND LOSS ACCOUNTS OF FOREST ENTERPRISE 'X', FOR THE YEAR ENDED . . . . 19 . . . . .

(after M. Gane)

	£		£
Forest production			
Afforestation and regeneration		Sales of standing timber	
Tending and maintenance of growing stock		Standing timber felled and converted by Enterprise 'x'	
Protection of the growing stock		Change in value of growing stock	
Balance, profit on forest production		Value of increment	
		less Value of removals	
		Net increase in value of growing stock	
Nurseries			
Raising seedlings and transplants		Sales of plants	
Decrease in stock of plants		Plants used in forest production	
Balance, profit on nursery operations			
Felling and conversion			
Standing timber from forest		Sales of timber	
Felling, extraction and conversion		Increase of stocks	
Carriage and other expenditure			
Balance, profit on felling and conversion			
Overheads			
Forest (holidays, supervision, etc.)		Other income	
Administration (salaries and expenses)		Profit on forest production	
Office (salaries and expenses)		Profit on nurseries	
Interest charges		Profit on felling and conversion	
Balance, net profit on Forest Enterprise 'x'			

increasing income becomes evident so that means to increase efficiency can be devised.

To attain that dual purpose it is necessary to use a method of accounting (and subsidiary cost accounting and analysis) which suits the kind of business so that costs and related income are distinguished according to the type of operation. The sources of costs and their relation to income and the general conduct of the enterprise can then be established.

Although principles and methods of accountancy cannot be elaborated here beyond their great value as a guide to improving efficiency when accounts are properly designed and used, samples of profit and loss accounts (after Gane, 1966) for a forest enterprise are shown in Table 29. It will be noted that operations are divided into several subsidiary enterprises and each of these may summarize regional subdivisions. Thus the Forest production might be regionally subdivided for two or more subordinate administrative charges. Other ancillary or subsidiary enterprises can easily be added, such as sawmills.

Budget. In effect the finance section of a working plan (in Part IV) should contain a budget which forecasts the changes in income and expenditure, growing stock, other assets and liabilities which will result from execution of the plan. To do this will require an operational budget summarizing the income, expenditure and profit (or loss) now and what they are expected to average during the period of the plan or, perhaps as usefully, what they are expected to be in the last year of the period. A budget balance sheet could also summarize the forecasted assets and claims. For details on the compilation of such budgets reference should be made to works on accountancy such as 'Business budgets and accounts' by H. C. Edey (1966).

Ideally such profit and loss budgets should apply to all facets of the forest enterprise but in a working plan it may be possible to apply them only to those facets of the enterprise which are dealt with comprehensively in the plan, e.g. the production forest and the felling and conversion.

The appendices

The appendices contain the basic data on which the working plan is founded and therefore form a bulky yet essential part of the plan.

Compartment histories. Some of the data, such as the compartment histories or registers may be so bulky and yet form so intergral a part of daily management that they may not be bound with the plan even as another volume but are kept in a separate series of volumes, probably in loose-leaf form and probably only in duplicate—one working, field copy kept by the sub-manager or forester and the other kept in the main office.

It is an essential and laborious duty for the working plan officer to bring up to date all compartment records including where necessary, as it is for selection forest working, graphs showing past and present curves of stem-frequencies by size-classes. It is also desirable to include a stock map in each compartment record, particularly if the compartment contains several different stands and sub-compartments in intensive management and also if the management is extensive and the compartments are large such as 500 acres or more. The compartment record form should contain space for entering details of treatment as it occurs (as stated earlier in this chapter) and also a sheet for inspection reports which *must* always be dated as well as signed by the inspector. In large extensively managed forests a compartment may be seldom visited unless it happens to be under regeneration. Inspection reports can lose most of their value if the time (year *and* month and preferably also the day) is not recorded.

Area statement

The compartment records form the base on which is compiled the detailed area statement which should correspond exactly with the legally recognized area of the forest. The statement should also show, compartment by compartment, the extent of administrative charges, silvicultural sub-division such as working circles and also areas not actually subjected to forest working, e.g. land leased for agriculture, but included in the working plan area.

Possible headings for an area statement with hypothetical entries are shown in Table 30.

Maps

Although maps to show several different features are usually essential it is a mistake to reduce the number of maps by combining their functions at the expense of clarity. Some maps, such as those for geology and soils, can often be combined without confusion of colours or symbols.

The stock map is usually the most difficult to keep clear even if it is on a large scale for intensive forestry. Although no immediate visual impression is thereby given the best way to

Table 30

AREA STATEMENT

Adminis-trative charge	Wood, block, enclosure Name	Area (acres)	Compt. No	Area (acres)	Areas allotted to working circles Conifer (acres)	Hard-wood (acres)	Amen-ity (acres)	Unallotted to WCs Area (acres)	Land Use
			1	34	24	6	—	4	Nursery
			2	28	20	5	—	3	Marsh-land
	Z	274	3	35	35	—	—	—	
			4	40	18	16	6	—	
A			5	36	—	—	12	24	Agric. fields
			.	.	.	.	.	.	.
			.	.	.	.	.	.	.
			.	.	.	.	.	.	.
			9	20	13	7	—	—	
	Y	340	10 to 20		and so on				
	and so on								
		2,640		2,640	1,566	889	140	45	
B					and so on				
TOTAL FOREST		15,635		15,635	12,326	2,650	568	91	

Note. (i) In this example the areas unallotted to working circles have been included in compartment areas and in the woods, block or enclosures.

(ii) Compartments are shown numbered consecutively. Frequently separate sets of serial numbers are provided for each containing block or wood.

record information on a stock map is usually by numerical and literal symbols rather than colours; shades of colour and cross-hatching become too confused to have a value in a complicated stocking under intensive management but may be the best way to distinguish broad forest types in extensive management. An example of symbols used in Sweden whereby, for a simple distribution of species, a stand or sub-compartment is adequately described on a map as shown below:

$$\frac{25-3-640}{\mathrm{III}-110-\mathrm{IV}}$$

where the symbols, reading from left to right, first above and then below the line, mean:

25—the serial number of the stand.

3—quality class, stating the MAI in cubic metres per hectare at an age of 100.

640—species, namely 60 per cent Scots pine, 40 per cent Norway spruce, 0 per cent hardwoods, only those three species or types being shown. Pure spruce would be recorded as 0100.

III—age class, e.g. in twenty-year intervals (I = 0–20 class).

110—standing volume in cubic metres per hectare.

IV—cutting class, of which there are five namely:

I—bare land and unestablished regeneration, II—established regeneration in the cleaning stage, III—middle-aged stands, IV—older stands not requiring regeneration, V—stands needing regeneration.

What is essential in a stock map is that the boundaries of each recorded variation in stocking should be clearly and accurately shown so that the area of each distinction can be measured on the map and found on the ground.

A method of symbols similar to those used in Sweden could easily be used elsewhere. For example a yield class in Britain could be shown as 22 to mean maximum MAI of 220. The main difficulty would be to show species in forests where numerous species occur. For that a light ground colour wash might be helpful—e.g. green for hardwoods, blue for spruces, yellow for firs and blue for pines but a mixture of such types (e.g. spruces and firs) would be troublesome and a method less exact than the Swedish one would have to be used such as SS/NS indicating a mixture of Sitka and Norway spruces of which Sitka is the more prevalent. The age however could be shown more closely than the Swedish method by giving the year of plantation or average natural regeneration, e.g. 25 or /86, meaning planted in 1925 or in 1886, the / indicating the previous century; two strokes would indicate the century before that.

The management map should show clearly the basic silvicultural management. That will include working circles and any allotment to periodic blocks, section and particularly the regeneration block or area. As periodic blocks and areas of regeneration, establishment and thinning affect daily management and locate the main types of treatment they are probably best shown in colour, as is done in France—e.g. *quartier bleu, jaune* or *blanc.* That implies that working circles should be shown by symbols—such as widely spaced light cross hatching or simply by lettering, e.g. CWC for the conifer working circle. Sections equivalent to the felling cycle in a selection system are best indicated by arabic numerals which may even be the year when fellings are prescribed, but sections corresponding with the thinning cycle are best omitted unless it can be done without confusing the map. If the final fellings are rigidly prescribed, as they might be for coppice or clear felling, the blue regeneration area can be divided into coupes allotted to the year prescribed for fellings; for more flexible felling arrows can show the recommended progress.

The management map can usually accommodate non-silvicultural programmes such as those for new construction or major reconstruction of roads, sites for projected buildings, timber collection depots or timber processing plant and sites for new villages or labour camps needed in remote, under-populated tracts.

Scale of maps. For intensive management a scale of 1/10,000 or 6 inches to one mile may be adequate but 12 inches to a mile may be necessary although larger scales than that are needed only for highly intensive work needing species treatment such as poplar cultivation. In India the standard scale, established before the end of the nineteenth century, was 4 inches to a mile and proved adequate. But in many tropical countries maps have only recently become available and the standard scale of the topographical map is often as small as 1/30,000 or less, or about 2 inches to one mile or less which is suitable only for very extensive forestry. The scale 1/30,000 or 1/40,000 is also too small for interpretation of aerial photographs except for distinguishing broad forest types. Generally for useful inter-pretation of aerial photographs scales greater than 1/15,000 and preferably 1/10,000 are needed (Rosayro, 1962). The importance of scale relative to the selected size of working unit is realized when it is noted that 1 acre is about 70×70 yards on the ground; on a map it is only about $0 \cdot 23 \times 0 \cdot 23$ inches at a scale of 6 inches to the mile, or less than $\frac{1}{4}$ inch square.

Contours are invaluable for road planning and site assessment and draining schemes. If they are not available arrows can be drawn on maps to indicate slope and aspect, the shorter the arrow the steeper the slope. But for the management map it is best to use a skeleton map showing only boundaries of the forest, compartments and other territorial units, roads and rides and watercourses. More detail is unnecessary and confuses the management entries.

Inventories

The appendices should include results of inventories and details, by example at least, of methods of sampling and calculations used. There should be a complete statement of the volume of the growing stock and its increment, the details of which will depend on the silvicultural methods used and intensity of management. The results of the inventory could be recorded compartment by compartment, summarized for each wood or block and for each working circle and felling series and if necessary separately for each main species and together. The form of record will naturally vary with circumstance but the compartment or sub-compartment will usually be the basic unit of record even if inventories have been based on sampling forest types rather than territorial units. What is essential is that the inventories and form of record should be planned to suit working methods.

Samples of the form of record used in Britain are given in Appendix I, in Forestry Commission forms WP. 6 and WP. 18 which are built up from the sub-compartment and compartment records in forms WP. 0 and WP. 1. In those forms increment is not used, nor are volumes in compartments and sub-compartments, but emphasis is laid on the assessed yield class, area, species and age, which are all that is necessary when there are reliable yield and production tables. The forms suit the method of working by area and thinning controls and production forecasts based on yield class, species and age.

For more mature forests where the final cut has more importance than it now has in Britain, particularly if final yields are based on volume and increment calculations related to normality of a sustained yield, it becomes essential to know at least the volumes and increment of the regeneration block and, for the regeneration area method or any method using a global yield, similar figures for the whole growing stock.

Glossaries

Glossaries should not be forgotten even if the plan is designed only for local use. It is often convenient to use colloquial or vernacular names of plants, trade names or even local units of measure which can then be extremely confusing for a newcomer to the forest and unintelligible outside the country if there is no glossary. For example, the cord is a widely recognized unit of measure for wood but even in England its actual dimensions may differ from place to place.

Bibliography

Anon. (1965) *Forestry Commission, Working Plans Code*

Bourne, R. (1934) 'Working Plan Headings', *Commonwealth Forestry Review*, Vol. 13 (2)

Bradley, R. T., Christie, J. M., Johnston, D. R. (1966) 'Forest Management tables', *Forestry Commission Booklet No 16*. HMSO, London

Brasnett, N. V. (1953) *Planned Management of Forests*. George Allen and Unwin, London

Edey, H. C. (1966) *Business Budgets and Accounts*. Hutchinson & Co., Ltd., London

Gane, M. (1966) 'The Valuation of Growing Stock Changes', *Quarterly Journal of Forestry*, LX (2)

Forestry Commission (1966) *Forty-Seventh Annual Report of the Forestry Commission*. HMSO, London

Hiley, W. E. (1954) *Woodland management*. Faber & Faber, London

Meyer, H. A., Recknagel, A. B., Stevenson, D. D., Bartoo, R. A. (1966) *Forest Management*, 2nd edition. The Ronald Press Coy., New York

Rosayro (1962) 'Ecological significance of first types', *Photogrammetria* XIX (1)

Chapter X

THE HISTORICAL DEVELOPMENT
OF FOREST MANAGEMENT

PREHISTORIC AND EARLY TIMES

Until he invented tools with which to cut and remove trees and until he discovered how to use fire to burn and destroy forests, man had no more impact on forests than any other animal. Even then man's impact on forests was negligible until his command of his environment so developed that his population increased materially and at the same time his wants diversified.

In those very early days man's attitude to the vast forests which enveloped him was presumably a mere acceptance of them as a fact of his environment, as a source of food and shelter enjoyed in the same way as he enjoyed water, light and air, warmth and cold. He accepted the forest as a vast and indestructible part of nature but entertained no concept of personal possession of it.

As man slowly learned how to control and use the products of his environment by clearing and cultivating the land, tending domesticated animals and building shelters against the weather and his enemies, his general attitude towards the forest was to deplete and destroy it to make room for his fields, flocks and herds. He also used the forest as a source of food from fruits, roots and hunting and of raw materials for building houses and stockades and for fuel to cook food and warm himself. In spite of this multiple use of forests, however, their vastness and apparent general indestructibility prevented man from conceiving ideas of possession until after he had cut most of them down and converted the land to tillage, pasture or waste. The forests themselves remained a gift of nature for all men to use alike as they wished and were subjected to unregulated multiple use by means of uncontrolled selective cutting.

It was not until man's population multiplied, the extent of forests diminished and their products become locally scarce that ideas of possessing forests began to arise concurrently with a realization of their value and, to some extent, their destructibility. But the realization was slow and tempered by the ability of forests to re-invade land that had been cleared and settled in agriculture as soon as the land was abandoned or neglected owing to war, pestilence and movement of population.

The very slow appreciation of the value of forests and the need to conserve and tend them is emphasized by the paucity of Biblical references to forests or woods. Early references refer to woods as a source of land for cultivation at the time of apportioning land to the tribes of Israel (Joshua 17, vv. 15, 18). References were also made to the value of timber in Lebanon for building Solomon's temple (II Chronicles 3, vv. 8, 9). But there does not seem to be any reference to the tending of woods, except of individual trees such as the olive or vine and references to the horror of fire (Ezekiel 15, v. 6).

The immense destruction of forests in Egyptian and Biblical times was exemplified by the Phoenicians who were the principal agents of destruction in Syria and Israel for their naval construction and sea commerce (Maury, 1850) but records of tending forests are hard to find until we consider Roman regulations in Italy. According to Maury (1850) and his references to Pliny, the Romans distinguished between coppice and high forest, regulated amounts cut and introduced penalties for cutting and damaging trees. Nevertheless the Romans were interested mainly in the increase of cultivated, agricultural lands and their forest management seems to have been confined to regulation of the rotation and annual cut in coppice forests and prevention of major abuses. Pliny recorded the rotations used in coppice, e.g. eight years for sweet chestnut and eleven years for oak.

The very first idea of preserving forests probably arose from the veneration held for large trees and dense groves of fine trees and their relation to religious and spiritual convictions. The size and majesty of individual trees and the mystery of a dense forest stimulate the belief that they provide a home for the gods and must accordingly be preserved as sacred groves. The sacred grove at Delphi in Greece is an example.

Nevertheless whether the first protection of forest was for religious or more material reasons, it is clear that no thought was given to the general care of forests until they began to become scarce in regions which were well-populated and where the social structure had been organized. Then claims of owner-ship and delineation of forest property, without which manage-ment is hardly possible, began whether the claimant was an individual or a community such as a village, town or religious body. Frequently ownership would be vested in the king or head of a tribe or people as a result of the custom that 'waste' land, i.e. land unoccupied by an individual, or even *all* land was owned by or vested in the king or chief. At the same time rights of usage of the products of the forests, as a relict of immemorial custom, without ownership of the land might continue provided that the owner's wants from the property were not affected. Thus it might be that the owner reserved to himself the great trees and the hunting (for its excitement but also for its great food value as protein) but allowed rights of common to arise for edible fruits, pasture, dead wood-fuel, brushwood and perhaps the less valuable tree species.

In that connection Maury (1850) records that in ancient France two uses of the forest were recognized, that of the chase which included fishing, and that of forest products. Also the *Capitulaire* of Charlemagne in the year A.D. 813 recognized two kinds of forest, those owned by the king and those owned by individuals. The French forest administration is still known as that of *Les Eaux et Forêts*.

Hunting and the development of forest policies in England

The influence of hunting on forest management in Europe has been so great that it is worth considering its effects in England where French practices were introduced by William the Conqueror and continued by his successors to the throne with long lasting results. But previous to 1066 King Canute had proclaimed similar protective laws in English forests. Stubbs (1913) points out that Canute's legislation was simple: 'I will that every man be worthy of his hunting in wood and field on his own estate. And let every man abstain from my hunting: look, wherever I will that it should be freed, under full penalty' (II Cnut, § 80). The practice of preserving the chase was established in England long before 1066.

The forest law to protect hunting as developed by William the Conqueror and his successors was more widespread than that of Canute and was severely administered by a body of chief foresters, foresters, and other officials and enforced at the Forest Eyre or itinerant court of justice. The forest law did not apply only to the woods of the king's domain and to that extent was similar to the universal protection given to a 'royal' tree such as the teak in Burma. To quote Turner (1901): 'In mediaeval England a forest was a definite tract of land within which a particular body of law was enforced, having for its object the preservation of certain animals *ferae naturae*. Most of the forests were the property of the Crown, but from time to time the kings alienated some of them to their subjects. Thus the forest of Pickering in Yorkshire and all those in the county of Lancaster were in the fourteenth century held by the Earls of Lancaster, who enforced the forest laws over them just as the king did in his own forests. But although the king or a subject might be seised of a forest, he was not necessarily seised of all the land which it comprised. Other persons might possess lands within the bounds of a forest, but were not allowed the right of hunting or of cutting trees in them at their own will.'

The Assize of Woodstock and succeeding Forest Charters

The provisions of the forest law up to the end of the thirteenth century were embodied in a number of charters of which the first was the Assize of the Forest, or the Assize of Woodstock, of 1184 in which Henry I incorporated the customs and practices of the law as they had been applied to that date. The great Charter of 1215, of John, which alleviated some of the previous rigours of the forest law, was succeeded by the Charter of the Forest, 1217, of Henry III, by which the forest law was fully exposed in a separate charter. That charter was itself succeeded by a re-issue in 1225, of Henry III, which introduced more alleviations such as the exclusion of penalties of life or limb for killing a deer and finally in 1297 a confirmation of the charter by Edward I.

In general the several charters after 1184 introduced successive alleviations of the forest law by disafforesting certain tracts, by reducing penalties to fines and imprisonment only, by stipulating enquiry into and by requiring punishment of corruption and oppression by forest officers, by reduction of the numbers of forest staff and by granting some allowances to barons and clergy to kill deer when in transit through forests on the king's business.

It is noteworthy that the forest law had only one avowed object, namely the preservation of hunting (primarily for the king) although the monetary income from fines and charges must have had substantial consequence.

It was fully appreciated that in order to attain the object the methods adopted should:

(i) Maintain or increase the population of the beasts of the forests or venison (four and only four species—the red and fallow deer, the roe and the wild boar (Turner, 1901)).

(ii) Preserve the free movement of the venison.

(iii) Preserve the habitat, i.e. forest structure, of the venison.

But the processes of accomplishment were never constructive but entirely negative by means of prohibitions and penalties, namely:

(a) No killing or possession of venison.

(b) No possession of unlawed dogs in the forest. (To law a dog entailed removal of three toes of a fore-foot.)

(c) No uprooting of trees and no reduction of the land to cultivation. (No assart.)

(d) No encroachment on the forest by enlargement of a curtilage (ground round a house) in the forest, no erection of a mill or fishpond, no enclosure of lands within a forest by a ditch or hedge without permission. (No purpresture.)

(e) No cutting of trees (without permission) beyond the rights of cutting wood for fuel and repair of property. (No waste.) (The rights varied and might include lopping, removal of brushwood or even trees other than oak, subject to supervision by foresters.)

During the fourteenth century enforcement of the forest law became more and more lax, but it was not until the end of the fifteenth century that a new and somewhat more positive forest act was proclaimed when conditions were very different. After the Black Death of 1349 population and prosperity multiplied. Chaotic labour conditions after 1349 had stimulated the wool trade which throve to the advantage of the yeoman farmer and industrialist. Progressively more land came under cultivation and more timber was felled for construction of ships and larger more luxurious houses built largely of wood. Wood and charcoal were the fuels. The forests were rapidly deteriorating and disappearing, a process that was accelerated by the lack of fences and consequent free range of pasture within the woods so that seedlings and coppice regrowth were destroyed.

The Acts of Edward IV, Henry VIII and Elizabeth I

The act of 1482, Cap. VII of Edward IV, was short and simple. It merely allowed owners of land within Forests, Chases and Purlieus to enclose their woods after felling with hedges against all beasts and cattle for a period of seven years to preserve the young growth.

But forest destruction continued apace and timber began to become scarce. At last a forest act that defined a positive policy appeared in 1543, Cap. XVII of Henry VIII, entitled 'The Bill for the Preservation of Woods'. Its provisions included compulsory enclosure of regeneration areas and silvicultural directions on the reservation of seed trees.

In summary the act distinguished coppice woods from high forest and woods subject to rights of common from unburdened woods. Both coppice woods and high forest had to be enclosed with fences after felling and standils or storers renewed in coppice and great trees in high forest (oak, failing which elm, ash, asp or beech) to the number of twelve each acre.[1] In coppice woods the standils could not be felled until they attained 10 inches square at 3 feet from the ground; in high forest they had to be kept for twenty years. The term of enclosure was seven years in high forest and four or six years in coppice depending on whether the coppice rotation was fourteen years or less or above fourteen years up to twenty-four years respectively.

In woods subject to rights of common the owner was not allowed to fell trees (except for his own use) until he had arranged with the commoners (settled if necessary by two justices of the peace) to enclose one quarter of the area which would be fenced; felling would then take place subject to the reservation of standards as elsewhere. The enclosure was valid for seven years.

In general, swine above ten weeks of age were prohibited

---

[1] The Act of Henry VIII distinguishes coppice having a rotation no more than twenty-four years from forest containing trees aged more than twenty-four years. The author here interprets the latter as high forest. The 'great' trees, therefore, *may* not be much older than twenty-four years.

from entering any enclosure unless ringed or pegged but colts and calves under one year could be put in enclosures two years after their felling.

Experience showed that the terms of enclosure were too short and they were extended by two years by the next act, Cap. XXV (§§ XVIII and XIX) of Elizabeth I in 1570 which also excluded calves and colts from enclosures for five years.

The development of forest management in France

Legislation to control the management of forests developed more quickly in France than in England. Coppice working, which had been practised from at least the times of Charlemagne in the early ninth century, was well understood in mediaeval times. In France as elsewhere the problems always to be solved were:

(i) The control of supplies to and the practice of right holders whose grazing cattle entered forests and destroyed coppice regrowth and seedlings.

(ii) The prevention of inevitable abuses wherever there were small scattered coupes or selective fellings in high forest where merchants could easily defraud the owner and corruption among the supervisory forest staff was easy and common.

(iii) Successful regeneration by seedlings in both coppice and high forest.

Consequently there was a tendency to adapt to high forest the understood method of coppice working which was used in all small forests and around the perimeter of big forests in whose hearts was the high forest maintained for special fellings by the owner. Consequently coupes in both coppice and, later, high forest, tended more and more to be laid out not only by area for clear or nearly clear-felling but also so that the successive coupes adjoined and touched each other and progressed in one direction whereby their progress could be surveyed easily and be seen to be done. The many ordinances in France, applying in the main to the forests of the royal domain, illustrate how the difficulties of management were met, difficulties which may still apply in some undeveloped parts of the world today.

The forest ordinances of France to the seventeenth century

Although there were earlier ordinances, e.g. those of 1280, 1318 and 1346, the most famous, thorough and long-lasting was the Ordinance of Mélun of 1376. The ordinances of 1280 and 1318 were the first to regulate fellings for right holders and prescribed the procedure for laying out the coupes while the ordinance of 1346 organized for the first time the administration of the forest (although 'masters of the forest' had been mentioned in 1318) and initiated sustained yields by requiring the 'masters of the forest to investigate and visit all the forests and organize the coupes so that the forests can perpetually sustain a good state' (translated from Huffel's (1926) quotation from the ordinance).

The Ordonnance de Mélun (Charles V), 1376

In effect this ordinance lasted for nearly three centuries until the ordinance of 1669 since its main prescriptions were repeated in the several subsequent ordinances (Huffel, 1926). Its main prescriptions were:

(i) Right holders would be allowed only the 'possibility' or suitable output of a forest. (This was the first reference to a yield regulated so that it can be furnished without diminution; it was more explicitly described or defined in 1669.)

(ii) Fellings in high forest were to be by area of 10–15 hectares, as in coppice, and not be numbers of trees as previously. The coupes were to be selected by the masters of the forest in assembly with their officers.

(iii) Coupes were to be clearly delineated and surveyed by lines that were distinguished along their length and at each corner by trees that were hammer-marked at their base and, later, at breast height also.

(iv) Coupes were to be enclosed by hedges or fences after felling.

(v) Standards were to be reserved in both coppice and high forest coupes to a minimum number of eight or ten each arpent (i.e. 6–8 the acre).

(vi) Trees should be felled with care to facilitate regeneration.

Subsequent ordinances did not change the basic provisions embodied in the ordinance of 1376 but rather made them more precise or added details such as provisions to regulate hunting or to stipulate that the trees marked and reserved along boundary lines and at corners should be selected from the best stems, preferably oak.

The tire et aire method

The crucial matter was the choice of the position of the coupe in high forests. It *should* be chosen in localities that need regeneration, namely where the trees are old and mature or deteriorating and that practice was prescribed in the sixteenth century. But to choose such areas for felling tended to cause considerable disorder by scattering successive coupes all over the forest. Consequently the practice did not last long and the ordinance of 1544 prescribed for high forest the practice long used in coppice namely that coupes should be selected *à tire et aire*, i.e. that the coupes of the year, selected by area, must have at least one boundary contiguous and coincident (à tire) with one boundary of the last coupe and that successive coupes proceed in one main direction. That procedure was made more explicit by the ordinance of 1597 which stipulated that coupes should be *de proche en proche, à tire et aire*. The need for this great care and formalized yet simple procedure becomes clear when one considers that the high forests were large, unmapped and without a network of roads and rides. To facilitate survey and maintain order in working the method *à tire et aire* was essential and provided an orderly sequence of age-gradations moving steadily round the forest to form compact blocks of age-classes which later enabled the periodic block methods of working to develop.

Those mediaeval developments which required all fellings to be approximate clear fellings in contiguous coupes, surveyed and enclosed, applied only to the extensive broad-leaved forests of France and not to the less accessible, mountainous coniferous forests such as those of the Jura and Vosges. The forests of the Vosges were managed by their owners for commerce (principally the dukes and abbeys) as well as for pasture, with a system of sawmills placed on the waterways for power and convenience of access. Supplies for each sawmill, perhaps 200 large trees yearly to a mill (Huffel, 1926), were cut selectively until the nearby forest was exhausted when the mill was moved elsewhere.

But Huffel points out that by experience the foresters learnt what area of forest was needed for the continuous support of a mill so that the practice of selection fellings with a sustained yield regulated by numbers of trees developed.

Thus we see that by the middle of the sixteenth century France had achieved, for its broad-leaved, royal forests at least, an organized forest administration and an ordered silvicultural management designed to prevent abuses by prescribing contiguous coupes, based on clear-fellings with reservation of standards, equal in area and proceeding in a definite direction only returning to the origin after a determined period (not less than 100 years in high forest or ten years in coppice).

In the sixteenth century also silvicultural techniques were advanced and thorough, namely:

(i) Retention of an adequate number of suitable mother trees for natural regeneration, not merely the prescribed minimum of six to eight each acre *plus* the reserved boundary trees.

(ii) The sowing of seed by hand after working the soil to supplement natural regeneration.

(iii) Conservation of advance growth by, for example, reserving all oak and beech less than 2 feet girth at 6 inches from the ground, e.g. at Compiègne.

(iv) Cleanings at ten-year intervals to favour oak and additional seed sowing or planting where necessary.

(v) 'Improvement' fellings after stands reached the pole stage by removal of deteriorating mother trees and 'wolf' trees and cutting out undesirable coppice shoots and inferior species.

Unfortunately from the latter half of the sixteenth century and particularly in the seventeenth century all the progress made was steadily lost owing to the failure of the human element and the return of abuses and corruption in a deteriorating administration. Rising prices and demand and consequent greater temptation allied to a general decline in moral standards at the time must have been the causes and resulted in the loss of the silvicultural knowledge gained and in the overcutting and destruction of the forests.

The reformation of Colbert and the Ordonnance de 1669

Huffel (1926) points out that, as in 1376, it was anxiety over the supplies of naval timber that stimulated reform of the bad state of the French forests and enquiry into and correction of the negligence and corruption that had rotted the forest administration. Colbert, Controller of Finance, was put in charge of the reformation and by Council's decree of October 1661 the forests were closed to all fellings for three months during which time the forest grand masters, supervised by special envoys, had to report on the area and stocking of the forests and on all fellings and alienations that had been made since 1635.

After dismissal and severe punishment of all the grand masters in 1664 and 1667 and after a period of edicts, instructions and regulations the Ordonnance de 1669 (Louis XIV) was issued. The ordinance which was a law to control forest policy reduced the number of grand masters, required a definite formal procedure for approval of all fellings and provided for regulating rights of usage but did not change the good practices that had been developed in the previous century. Its object was to prevent abuse and malpractice in order to allow healthy forest management to develop.

It prescribed a tight control of the extent and siting of annual coupes by requiring each grand master to approve them after formal inspection and according to the accepted management in force or to authorize after justification any extraordinary coupe that was judged to be necessary. The formal procedure imposed continues to this day (Huffel, 1926). Beyond specifying the minimum number (eight to the acre) of seed bearers or standards in addition to the reserved boundary trees to be left after felling, the ordinance did not control silviculture by, for example, prohibiting thinnings or improvement or selections fellings and by requiring fellings to be nearly clear fellings. Nor did it specify that successive coupes should be contiguous; but they had to be surveyed. But the condemnation and penalties for abuses led in practice to the discouragement of thinnings and improvement fellings which by their nature encourage abuse. Consequently there was a deterioration in silvicultural practices which was not corrected for some time. But as the ordinance required all exploitation in royal forests to conform with decrees of Council it was always possible to change, adapt and improve silvicultural management. So from the early eighteenth century varied types of silvicultural management developed as the result of special regulations.

EUROPEAN DEVELOPMENTS FROM 1700 TO 1825

Hitherto progress in European forestry had been mainly in France. In Britain there had been and continued to be no advance. Forestry consisted merely in felling woods and periodic, feverish spasms of planting when supplies of timber ran short especially during and after wars. Examples are seen in the writings of John Evelyn especially in Sylva published in 1664 advocating afforestation and care of woods, and the plantings made at the end of the seventeenth century and again from 1805 to 1820 notably in the New and Dean forests. Legislation authorizing new enclosures (e.g. in 1698 and 1808 for the New Forest) was passed by Parliament.

The next advance in European forestry, chiefly at the end of of the eighteenth and beginning of the nineteenth centuries when France was heavily engaged in political affairs, was in Germany and Austria where basic thinking showed the way to great developments and saw the introduction of schools of forestry such as G. L. Hartig's school in Hessen in 1789 and H. Cotta's Saxon school in Thuringia at Zillbach in 1795. The forest school of Nancy in France was founded in 1824 by Lorentz who had studied under Hartig.

It must be remembered that throughout the period from 1700 to, say, 1825 European forestry continued to be influenced by the slow methods of transport and the poor roads so that each small district had to be largely self-sufficient. Nor was this basic factor quickly changed by the introduction of the steam engine, the spread of railways, canals and ocean steam ships or by the discovery and use of alternative sources of fuel and power. It was not until the internal combustion engine became reliable and its use universal (after say 1925 or even 1930) and until small, local firms began to be replaced by large ones that the self-sufficiency of small districts began to be unnecessary.

Consequently, as we have seen the need for sustained yields from a forest had long been recognized but the means for attaining it had not advanced beyond the use of a rotation and an area yield for which a clear or near clear-cutting system seemed to be essential in both coppice and high forest. Only in some coniferous forests, as in the Vosges, had it been found possible (and then only by experience and not by calculation) to sustain the yield by numbers of trees cut by selection.

German developments

Huffel (1926) records that the first forester to have the idea of a volume yield was Bollem, in about 1740, when he was in the service of the duchy of Eisenach. Moreover he divided the rotation into a number of periods (e.g. eight periods in eighty years) and allotted to each period such an area of forest that the volume yield was kept constant throughout the rotation. Then in 1795 G. L. Hartig (1764 to 1837), who was later head of the Prussian forest service, co-ordinated the work of his predecessors and announced the method of working by periodic blocks with equalized volume yields although the procedure that he advocated was complicated. Compartments were allotted to their appropriate periodic block basically and firstly according to age. Then the yields from each compartment and hence by sum the yields from each block were calculated according to their initial volume *plus* their increment up to the middle of the period in which the crop matured (and was regenerated). If the yields so estimated (using yield tables) were found to be inconstant from period to period compartments were shifted from one period to another to equalize yields. This process of shifting naturally required a recalculation of the yield from the shifted compartments. As thinnings were also included in yields the calculations become very laborious. Moreover calculations and estimates involved long periods of time (up to rotation length less half a period) and were proportionally unreliable. So Hartig's method of allotment was not long-lived but his use of periodic blocks has continued to this day.

Later in 1820 and 1832, H. von Cotta (1763 to 1844) emphasized the essential factor of area and the need to simplify Hartig's use of volume, increment and equated yields for the allotment of compartments to periodic blocks. Cotta introduced the concept of reduced or equiproductive areas according to their productivity. Having allotted periodic blocks by age and equiproductive areas the final yield by volume from the regeneration block was taken to be the initial volume plus its increment for half the period, i.e. $AY = V + \frac{1}{2}I/P$, where V is the initial volume I the increment of that volume for the whole period and P is the period in years and AY is the annual yield.

Brasnett (1953) records that a Thuringian forester, Öttelt had shown still earlier (in 1765) how thinnings by providing early returns increased the profit gained from a crop as their income counterbalanced the effect of compound interest on the cost of formation. Moreover Öttelt was the first to express the volume of the growing stock in terms of the MAI (i) per unit of area of the final stand, so that $i = V/R$ where V is the volume per unit of area of the *final stand* and R its age or rotation. Consequently $V = i \times R$. Thence he showed that the volume of a complete series of age gradations in the middle of the growing season $= VR/2$. In this calculation however thinnings are *not* considered. Perhaps Öttelt's work was known to the tax department in Austria (see below).

Austrian developments

As described earlier in chapter VIII, p. 188, the Austrian Cameral valuation method for assessing the value of forests for taxation was introduced by decree in 1788. This method established the principle that a properly managed forest should have a series of age gradations to give a constant yearly yield in volume and income. That volume yield must equal the mean annual increment of the forest (excluding the complications of thinnings) and the standing volume of the forest must equal half that increment multiplied by the rotation (as shown by Öttelt).

Consequently it became clear that by cutting a volume equal to the increment a 'normal' forest would be maintained and that by cutting more or less than the increment (whichever might be appropriate) an abnormal forest could be led to its normal state. (The Austrian formula, for which see chapter VIII.)

Silvicultural progress in Germany

Lastly let us note the silvicultural development in Germany which were exemplified by two early ordinances. The first of these, the Hesse-Kassel Ordinance of 1711, prescribed the retention of about forty seed-bearers to the acre (ten to twelve paces apart) at the time of regeneration. The retained mother-trees were to be removed later as sufficient natural regeneration became established. The second was the Ordinance of 1736 in Hesse-Nassau in which three successive regeneration fellings, a 'shade', 'light' and a final felling, were prescribed similar to the several seeding, secondary and final fellings of the Uniform system of a later day and which are still practised today. Troup (1952) states that G. L. Hartig in his *Anweisung zur Holzzucht* published in 1791 recommended the same system; the 'shade fellings' left seedbearers almost touching and the 'light' fellings left seedbearers so that there was a gap of fifteen to twenty paces between crowns; the final felling was made when the regeneration was two to four feet high. Cotta later recommended the same procedure. As artificial sowing and planting had already been used earlier in Germany it is probable that inadequate natural regeneration was artificially supplemented where necessary. In the design of fellings we can see a contra-distinction and departure from the methods practised in the *tire et aire* of France in which the reserved standards were left to grow through the next rotation.[1]

Basic progress attained by 1825

By about 1825 the basic principles of forestry had been established in Europe, namely:

Before 1700. (1) The need for a sustained yield which implied a co-ordination of the suitable rotation with the area of forest cut yearly.

(2) The need to regulate enjoyment and satisfaction of rights and to protect regeneration. This required delineation and legal definition of the forest estate.

(3) The need so to organize and control fellings that opportunities for overfelling owing to miscalculation or human greed are removed.

From 1700 to 1825. (4) The use of periods and periodic blocks instead of the yearly coupe to organize work in high forest.

(5) The use of volume and not only area as a measurement of yield and therefore also a means to regulate the amount cut.

(6) Recognition of the basic properties of an ideally managed or normal forest and the consequent means for guiding or moulding a forest to normality.

[1] Troup (1952) records that Tristan de Rostoing, one of the grand masters and head of the French forest service, recommended as early as 1520 a form of successive regeneration fellings to replace the fellings of the *tire et aire* method of working.

11

(7) The effect of compound interest on profits.

(8) The introduction of successive regeneration fellings for the natural regeneration of uniform, even-aged stands.

It is evident that by 1825 continental European foresters, at least those in Germany, Austria and France, had grasped the basic principles of forest management. They understood or at the least had explored the relations between yield and increment, the fundamental properties of the even-aged, normal forest designed for sustained yields, the scope for regulating the cut by volume as well as by area and the effect of compound interest on profits. Long experience of forest destruction had previously taught them the need for legal definition and constitution of the forest for the proper protection of the forest estate in which work had also to be organized in a practical manner. Consequently forests were surveyed and demarcated and divided into felling series, or sustained yield units, which were themselves divided into compartments that were aggregated to form periodic blocks suited to the rotation.

The future must have been an exciting prospect particularly as markets must then have seemed secure in an economy which was generally expanding without undue instability caused by rapid, inventive progress or change in the standards and way of life. In which main direction would the next move of forest management be—towards the silvicultural variety made possible by yields flexibly regulated by volume instead of area or towards a more rigid, normal forest planned for ease of utilization and the best financial yields?

EUROPEAN DEVELOPMENTS FROM 1825 TO 1925

Even-aged forestry and periodic blocks

The first impact of the new forestry spreading from the forestry schools of Prussia and Saxony was the success of the uniform shelterwood system of regeneration and the use of periods and periodic blocks which were advocated and practised first by G. L. Hartig and then Heinrich von Cotta. A fear that wind-storms would upset both the mother-trees and the method was expressed in 1796 by the Bavarian forester Schilcher who proposed the retention of unfelled strips between successive regeneration coupes. Cotta improved on this use of severance cuttings by dividing the regeneration block into cutting sections in which fellings were confined yearly in turn to avoid adjacent coupes in successive years.

The influence of Hartig and Cotta was not confined to central Europe but spread to France where seeding fellings were started in the forest of Bellême in 1822. The great Nancy school of forestry was founded in 1824 under Bernard Lorentz (1775 to 1865) as Director from 1824 to 1834 who had been a pupil of Hartig. A. L. F. Parade-Soubeirol (1802 to 1864) who had studied under Cotta became a lecturer at Nancy in 1825 and Director from 1838 to 1864. The success achieved at Bellême and the advice and teaching of Lorentz and Parade encouraged the wide use of the Uniform system so that by the middle of the century it had ousted the old *Tire et Aire* in hardwood forests and was extending into the hilly coniferous forests of the Vosges and Jura. It remains one of the main systems of silviculture in France today.

As pointed out by Huffel (1926) the volumetric complications used by Hartig were repugnant to the French with their love of practical clarity so that it was the area methods used by Cotta for the allotment of periodic blocks that were developed in France. Although Cotta had never proposed any permanency for periodic blocks but rather recommended the frequent revision of plans, the advantages of permanent, compact periodic blocks in which different types of silviculture and utilization could be severally concentrated and which led the forest to normality in one rotation, greatly appealed to the French. Moreover the inevitable sacrifices involved in the formation and maintenance of a permanent pattern of periodic blocks were comparatively small in forests that had been well managed under the *tire et aire* and had attained a satisfactory series of contiguous age-gradations.

Introduction of more flexible methods

The permanent periodic block method with its rigid time-table was, therefore, widely and enthusiastically developed in France even in hilly coniferous forests where severe storms were comparatively frequent. Its use was relatively undisputed until about 1880 or 1890 but is rarely used today. The drawbacks to so rigid a method, outlined in chapter VI of this work, were increasingly recognized particularly in mountainous coniferous forests. But the French continued to recognize the advantages of even-aged forestry so that ways and means introduced to increase flexibility always retained the policy of growing stands that were as uniform as conditions of natural regeneration, windfall or insect attack would allow. Nor did French foresters, whose administrative charges have always been comparatively large, favour the intricate patterns of strip fellings (particularly in spruce forests) which developed elsewhere nor the single tree selection method and its intensive working.

Consequently, under the influence of Mélard (1842 to 1909), the French volume method of 1883 and the resultant regeneration area method which needed no definite time for completing regeneration but required a global volume yield embracing both thinnings and final fellings, began to be introduced into mountain forests instead of the fixed periodic blocks. Similarly the revocable and single periodic blocks, which did not require blocks to be compact and self-contained but allowed re-allotment at the time of revising plans, began to replace permanent blocks elsewhere. All these methods enabled rotations to be varied from compartment to compartment, but all aimed at maintaining uniformity of stands as far as possible. (For more details see chapter VIII.)

Conversion of coppice to high forest

French forestry in the ninetenth century was also greatly affected by the conversion of coppice with standards to high forest. Lorentz affirmed that to grow standards over coppice was a vice; woods should be either simple coppice or high forest. He also foresaw the coming fall in demand for wood fuel owing to the competition from coal. Vast areas of coppice with standards were accordingly put under conversion to high forest. Unfortunately Lorentz's method of conversion was simply to allow the coppice to age, subject only to light thinning, and merge with the canopy of the standards until it could be regenerated by the Uniform system. The consequent inevitable, sudden and heavy fall in yields and supplies to industry caused so much opposition that Lorentz was forced to resign in 1839. Parade and Nanquette replaced the method of conversion by the *Méthode classique* (as it has come to be known) by which the area was divided into four periodic blocks for a conversion rotation of some 120 years. Conversion began in only one periodic block which underwent a stage of waiting and ageing for the first period before regeneration by the Uniform system in the second period. That process was repeated successively in turn for the other periodic blocks which were subjected to coppice fellings until their time for waiting and conversion came. Parade's method (not introduced until 1873) therefore entailed a gradual fall in coppice yields but a longer conversion period. Today there are still large areas of coppice with standards in France and the need for their conversion is even more urgent economically than it was earlier. But now their conversion to coniferous rather than hardwood high forest may be more desirable.

Clear felling and planting pure stands

In spite of their preference for even-aged forestry French foresters did not clear-fell and plant on any large scale. Nevertheless large-scale artificial afforestation combined with land improvement, reclamation and stabilization was vigorously pursued with great success in the Landes along the coasts of Gascony in south western France. The great work, over 800,000 acres, was begun as early as 1789 at the direction of the Minister, M. Necker, but mainly after 1817 and completed in 1864. The pioneer work and particularly the techniques used to fix and afforest the sand dunes were by the famous M. Brémontier. Although the forests are worked by clear-felling of large coupes regeneration is natural from seed on felled trees or seed already on the ground.

In Germany, however, clear-felling and planting was used on a large scale. Cotta applied the shelterwood system to the better and more fully stocked hardwood forests where management and natural regeneration of the existing species were comparatively straightforward and satisfactory. But in the more mismanaged, degraded and open forests he introduced clear-felling and conversion to Norway spruce with fellings made in cutting sections to proceed against the direction of dangerously high winds. The certainty of quick success from artificial regeneration and the concentration of the final yield in a clear-felled coupe were as attractive then as now. In addition the market prospects for spruce were good and its volume production high. Consequently the practice spread rapidly in Saxony and indeed in many other parts of Germany but, unfortunately, the plantations of pure spruce were extended outside the true, hilly and damp habitat of spruce to the lower elevations and plains where the yearly rainfall was as low as twenty to twenty-five inches. Only in the upper zones (above 600 metres elevation) where rainfall is plentiful does spruce naturally thrive, probably mixed with silver fir and beech.

As early as 1820 Cotta warned against the indiscriminate use of Norway spruce and again in 1844. His warnings were disregarded. On the other hand Pressler, lured by the prospects of high financial yields promised by short rotations to supply pit-props to the expanding coal industry and presented with the fact of substantial and apparently successful young plantations, advocated the large-scale extension of pure spruce. Early success of the work in Saxony, the fashionable acceptance of and enthusiasm for normal even-aged forests managed to satisfy the theory of maximum soil rental (Faustmann's famous formula was published in 1849) and the simplicity of working pure crops encouraged others to copy Saxony. Clear-felling and planting with spruce spread to Bavaria, Bohemia, Austria and even to Switzerland where many forests had been depleted in the eighteenth century and whose foresters recruited after about 1820 were all German trained.

The subsequent development of many spruce plantations in Saxony belied early promise. Frost, snow and wind exacted their toll from the even-aged stands; fungus and insect attacks increased. But even more seriously soil fertility was affected. The deterioration was progressive and exemplified by unhealthy trees with short, discoloured needles. Raw humus accumulated and the ground flora changed from grass, blackberry, raspberry, and *Epilobium* to heather. Hard pans were alleged to appear in the soil and increment of the stands fell away.

The poor results of pure spruce plantations in Saxony were thoroughly investigated in 1921 by Wiedemann who in 1923 and 1924 published his results which Troup (1952) has summarized. Conclusions that could be made were that the adverse results were caused in part only by the clear cutting system as to some extent they also occurred in some woods raised under a shelterwood. But the poor results were intensified and probably in no small measure caused by the use of pure spruce instead of the original mixed crops and by the introduction of spruce into regions outside its habitat where the rainfall was too low. Wiedemann also concluded that the short financial rotations used in Saxony should be lengthened. One may also moralize that the rigid and large scale application of financial and mathematical calculations based on early performance and a constant productivity of a site for an indefinite time is unwise; harmony of species, site and treatment is fundamental. If special treatment (such as application of fertilizers) is found necessary to adjust site conditions the original calculations are likely to be upset.

Mixtures and irregularity

European foresters became anxious over the use of both pure species and even-aged stands long before 1921. Their anxieties were expressed and the courses and remedies studied and approached most notably by the great forester Karl Gayer in his comprehensive and penetrating book *Der Waldbau* published in 1880 and in other writings. He recognized that to provide sustained yields it is essential for the forest to be kept in a state of continuous stability so that production is not curtailed. To ensure the continuous productive capacity of a forest all the forces contributing to growth must be kept in harmony. Any treatment of the growing stock which disturbs and reduces the fertility of the soil must be harmful. Accordingly Gayer advocated mixed and irregular stands and the use of species climatically suited to the site. His thinking was guided by the perpetuity and health of the natural forest; although he emphasized the need to learn from nature and therefore favoured mixture and irregularity his approach was not impractical. By observance of nature and the irregular forest the forester can learn to tend stands from early youth to maintain a harmonious ecology.

In this brief historical account it is not possible to summarize or even draw conclusions from the works of Gayer and other foresters such as Gurnaud, Engler, Biolley, Schädelin and many others. Other works particularly those on silviculture by Köstler (1956) and Troup (1952) should be consulted. Here only the trend of thought and action is being considered.

It is clear that towards the end of the nineteenth century there was a resurgence of silvicultural thinking after a period of great managerial and economic development of the forest estate and study of the relations between growing stock volumes, increment, prices and rates of interest. But those very developments and their relegation of silviculture to a secondary role created the opportunity and spur to assess and critically analyse the results of the massed coniferous forests created since 1800 and to compare them with forests not yet converted to the then fashionable state. Consequently it was increasingly realized that silviculture should be based on the ecology of the forest as a vegetative community whose dynamic health depended on a harmonious relation between themselves and the soil which supported them. Interference among some members of the community might destroy the productive capacity of the whole. So there was a move towards the natural, irregular mixed forest; studies of the increment of classes of trees or even of individual trees began to be considered instead of the application of yield tables to the masses of a uniform and pure stand.

That is not to say that even-aged stands, particularly pure even-aged stands of Norway spruce were rejected. The controversy was only begun and in fact at the end of the nineteenth century the steadily rising demands for coniferous pulp-wood added to the old attraction for spruce. In the Ardennes of Belgium, for example, extensive plantations of spruce were raised whose conversion to mixed stands has been exercising the ingenuity of Belgian foresters in recent years.

Multifarious silvicultural systems

The keen silvicultural awareness of European foresters after say 1875 was shown in the design of silvicultural systems created to suit particular local conditions and solve practical problems of timber extraction or reduce wind damage. The systems so created have become the classic European silvicultural systems described so ably by Troup (1952); they constitute a memorial to the devotion, powers of observation and dedication and application of foresters who lived in the forests that they managed.

Among those systems was the Group system, developed if not invented by Gayer, which departed from uniform methods of regeneration by opening and then enlarging individual groups by peripheral successive regeneration fellings. In choosing the position of groups full advantage of existing gaps in the canopy and existing regeneration advance growth is taken. But in spite of his support of irregular stands Gayer was prepared to use short regeneration periods such as fifteen years although periods up to forty years were recognized.

Other systems introduced were variants of edge-cutting by which the regeneration coupe, subject perhaps to clear-felling but usually to successive regeneration fellings, is given the form of a narrow strip whose width varied from about half to the whole height of mature trees. The strips would be aligned at right angles to the main wind direction or the mid-day sun if desiccation was feared and successive fellings would proceed towards the wind or the southerly sun. Cutting sections were used to avoid having fellings in adjacent strips each year. There is no doubt that in many conditions the side shade provided by strip fellings favours natural regeneration and its development, particularly of Norway spruce which is relatively light demanding but sensitive to late frost. That advantage of suitable light conditions is increased by the narrowness of the strips and the gradual progress of shelterwood fellings which together prevent a sudden and drastic change in the rate of transpiration so that the soil water regime is largely unaffected. So silviculturally strips are admirable. Moreover they can be arranged to combat wind damage and avoid damage by timber extraction through young growth. There are however objections of which two are serious namely:

(1) Complication of design especially in hilly country. The design is also rigid and difficult to change subsequently.

(2) The speed of advance of regeneration is slow. Thus if strips are 30 yards wide and if regeneration is fairly quick and the next strip is cut after five years the speed of advance averages only 6 yards a year. If the strip is 200 yards long it will take one year to regenerate about ¼ acre or a hundred years to regenerate a compartment of 25 acres. If the rotation is a hundred years either every compartment must be under regeneration and contain all age-classes or each mature compartment must be divided into some four cutting sections.

C. Wagner in his *Blendersaumschlag* system introduced in Gaildorf in 1902 took advantage of using several cutting sections in a compartment by staggering the sites of the start of each series of strips which were *very* narrow and proceeded in a southerly direction to give protection from both the sun and a westerly wind as shown in Fig. 13.

Somewhat intricate variations are possible by using a diagonal advance in a south westerly direction and a variable arrangement in hilly country.

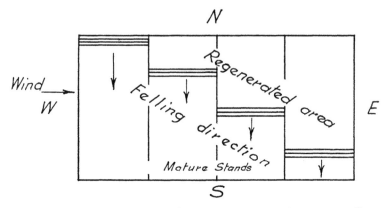

Diagrammatic representation of Wagner's Blendersaumschlag

FIG. 13

To increase the speed of advance and yet retain the advantages of edge-cutting the Wedge system (*Keilschirmschlag*) was devised by Eifert in 1903 and applied in the Langenbrand by Eberhard and elaborated by Karl Phillip in the Black forest of Baden where it is still successfully used locally on a substantial scale. The initial very narrow strip, pointed towards the prevailing wind, is enlarged by successive strips on each side and the tip of the original strip is progressively extended so that a wedge, steadily widened on each side is formed. The speed of advance of regeneration is thus doubled. The different light conditions in the strips on each side of the wedge may cause some difficulties in regeneration but encourage a mixture of species (e.g. light demanders on the northern side of the wedge and shade tolerant species on the southern edge).

Progress in Switzerland

In Switzerland the ideas of Gurnaud in 1879 were eagerly accepted and developed by M. Biolley to organize the selection system both silviculturally and managerially and to convert to irregularity the even-aged plantations and methods which were proving unsuitable even in the less mountainous forests of the country. In consequence the *methode du controle* was developed to measure, record and compare frequently the changing volume and increment of the growing stock in its various size classes for each unit of working. The method of control, which has been described in chapter VIII, made it possible to practice an intensive silviculture based not only on personal skill but on measured facts so that the selection system has attained a peak of perfection in the silver fir, spruce and beech forests of Switzerland.

In addition Switzerland, in company with Bavaria and Baden in Germany, has developed and is still developing *femelschlag* or the irregular shelterwood system. *Femelschlag* in Switzerland is really an amalgam of systems since by it different parts of a forest or felling series can be subjected to different types of treatment such as uniform, group, wedge or strip or even selective fellings as may be most suitable although clear fellings are always avoided. Overall the careful and intensive silvicultural tending of all stands from extreme youth to mature age is considered essential. Artificial regeneration is by no means excluded. The emphasis is on flexibility so that no rotation is used, the regeneration period is long but indefinite and the removal of stands or trees depends on their vigour and form rather than age. Much therefore depends on the skill of the forester who, aided by successive inventories and a close knowledge of volume, increment and yields of small working units, sets his own stamp on the form and constitution of the forest that he manages. But as the classic method of control is based on repeated measurements at breast height only it may not be truly applicable to stands, however small, that are essentially even-aged. Developments in the regulation of the cut and general organization of the very flexible system of *femelschlag* are still in progress.

Developments in yield regulation

The nineteenth century was marked by quantitative study of the volume and increment of the growing stock and its relation to the normal growing stock. Until late in the century the only normality considered was that of a forest or felling series composed of even-aged stands. Not only was such a forest the simplest model to consider but foresters genuinely believed it to be both the silvicultural and managerial best. Did not the dense even-aged stands copy the dense even-aged stands (though perhaps individually small in area) of nature? Let the natural thinnings of suppression by competition also be copied by very light artificial thinnings so that the fine timber quality and high volume (though at an old age) of natural forests could be reproduced artificially.

Much work was therefore done on the continent of Europe on the compilation of yield tables from data collected in permanent sample plots, in the stem analysis of trees and in the production of volume tables. Nor was the increment of standing trees neglected. Pressler (1868) introduced his increment borer which is still an essential tool today and showed how radial increment could be used to estimate volume increment.

The primary aim was to mould a forest into its normal state as expeditiously as possible by fixing a rotation, such as the financial rotation or that for maximum net income or volume, and by carefully regulating fellings. The periodic block methods using area were one way of regulation. The other was to regulate the volume cut so that the forestry of the nineteenth century was marked by the several formulae proposed—the Austrian formula and its modifications, those of Hundeshagen and von Mantel (and Masson of France) and later Mélard. Those formulae, which all used the basic model of the normal even-aged forest and have been used and modified throughout the world, have already been described in chapter VIII.

Stand management

No account of European forest management can omit a reference to the contribution of F. Judeich, Professor of forestry at Tharandt from 1866 to 1894, in his *Bestandswirtschaft* or individual stand management. Judeich in 1862 consolidated and made precise practices that had been developing since 1811 in the time of Cotta. Judeich's stand management aimed at coordinating the silvicultural and other attributes of every stand with the felling of stands to obtain a normal forest. He aimed to rationalize the process by mitigating the sacrifices imposed by a rigid selection of stands for final felling according to their age. As explained in chapter VIII Judeich's proposals required firstly a selection of stands which—for reasons of their silvicultural conditions perhaps owing to wind or other damage, or owing to over-maturity or other needs of utilization, or owing to their positions relative to other stands—ought to be felled and regenerated in the near future. Secondly the actual amount so selected would be modified according to the rate of adjustment deemed necessary to correct the proportion of age-classes.

In effect, therefore, Judeich's stand management was similar to the single periodic block method that was developed later in France. But his common-sense principle of basic selection of stands for final felling on their silvicultural health, degree of stocking and general state and performance in relation to the objects of management rather than mere age and the need to acquire normality, can and should be applied to any even-aged system of management.

Developments in Britain

It might be asserted that there was no progress in forestry in Britain in the period 1825 to 1925 until the publication in 1919 of the findings of the Acland committee and the passing of the Forestry Act of 1919 (9 & 10 Geo. 5. c. 58) by which the Forestry Commissioners and a State forest service were established. Only then was a stated forest policy combined with the means of its proper execution. It was the first time that that combination had been achieved since mediaeval times when the negative policy of preserving (by prohibitions) forests for the purpose of hunting had been supported by a recognized forest staff (albeit largely unpaid) which has been described earlier in this chapter. In the intervening centuries, except for the positive policy of protection and regeneration embodied in the acts of 1482, 1543 and 1570 there had been no real national forest policy nor staff to implement it. Crown (and to some extent private) forests were managed as individual forest estates by foresters and other staff appertaining to each forest to meet current demands and needs of preservation enlivened by spasmodic bouts of planting. There had been no continuous policy. But now a positive policy of afforestation and silvicultural improvement over all the forest land to provide a national stock of timber managed to provide continuously a truly useful increment and yield was at least announced and pursued.

There can be little doubt that the replacement of oak by iron and steel for building all fighting ships of the Royal Navy after

1862 (see reference to the battle of Hampton Roads on p. 337 below), the subsequent progressive disappearance of the wooden ship from the mercantile marine also and the ability to import timber cheaply had a disastrous effect on British forestry in the latter part of the nineteenth century. Hitherto British forest policy, such as it was, had in essence depended on the inborn conviction that ships of the Royal Navy at least must be built of English oak of which there had to be adequate supplies. Although periods of peace had always diluted enthusiasm the spasmodic but strenuous planting programmes at intervals of some 100 to 150 years and the high prices obtained for naval oak timber [1] had maintained some degree of continuing interest among both private owners and the Crown to grow and conserve oak forests. But after 1862 that spur to a forest policy had been broken only to be renewed by the 1914–18 war when the very existence of the nation was imperilled by the need to import timber, even at the expense of food and other commodities, in a mercantile marine decimated by enemy action. Consequently forests from 1862 until 1919 were neglected except by many private owners for the supply mainly of agricultural and estate produce, for recreation and for beauty. In witness of the apathy is the lack of age-classes planted between 1865 and 1915 in the New and Dean forests and elsewhere.

The development of forestry education in Britain

But the initiation of a forest policy and the means of its execution in 1919 was not the only achievement in the period under review. The training and education of higher forest staff began in 1885. It is, however, true that education of higher forest staff began in 1885. It is, however, true that education was directed entirely to training foresters for employment not in Britain but overseas in the Imperial Forest Service (later called

[1] According to Albion (1926) the Admiralty in 1804 paid up to £12 (or even £15 for special pieces) for a load of 50 cubic feet of oak timber, i.e. up to about 5s a cubic foot. The rise in average costs of oak timber to the Admiralty had been continuous, thus: 10s a load in 1547, £1 in 1588, £2 at the Restoration, £3 6s in 1740 over £7 in 1804 and about £13 in 1813. The amount of timber needed for new construction and repair of ships was by no means negligible. Thus, to quote Albion again, in 1790 (a typical year of peace) the mercantile marine used 80,000 loads or 400,000 cubic feet and the Navy used 25,000 loads or 125,000 cubic feet or 525,000 cubic feet in all in one year.

the Indian Forest Service) for service in India and Burma. In this notable innovation Britain owed her forest emancipation to a German forester, Dr D. Brandis, who was trained in Germany and was appointed by the Governor General of India, Lord Dalhousie, to the charge of the Pegu forests in Burma. Subsequently Brandis (later Sir Dietrich) was appointed in 1863 to be Inspector-General of Forests of India and Burma and in 1866 were recruited W. Schlich and B. Ribbentrop, both German foresters trained in Germany, for the first two appointments to the Imperial Forest Service. Until 1886 all those recruited to the I.F.S. had been trained either in Germany or at Nancy in France. Dr (later Sir William) Schlich, retiring from his post of Inspector General in 1889, founded and was Director of the first forest school in England in 1885 at Cooper's Hill near Windsor where there was already a school of engineering for the Indian Public Works Department. In 1905 the school of forestry was transferred (still under Schlich) to the University of Oxford where training was no longer only for those destined to serve in India. The school of forestry at Edinburgh University had already been founded in 1889 and later schools were established at the University of Aberdeen (1900) and at Bangor (1904), University College of North Wales. The chairs of all those schools were at first held by men retired from the I.F.S. so that the influence of the German and French schools of forestry and of Indian methods of management were substantial until 1925 when a more individual national approach to forestry began to develop in phase with the vast afforestation programmes and organization of British forests.

The influence of private forestry

In other respects progress in British forestry from 1825 to 1925 was confined largely to the unco-ordinated efforts of private individuals who owned the great majority of the comparatively small area of forests and woods. But their influence should not be ignored. Although none introduced new methods of silvicultural management, many were devoted to their woods and were active in introducing new and exotic species into their gardens, parks and woods and were pioneers in afforesting peat land so that by 1925 there were perhaps a greater diversity and quantity of exotic tree species growing in Britain than in any other country. Those early introductions provided material and ideas for expansion in the subsequent use of exotics in Britain on a scale greater and more varied than elsewhere thereby starting a trend which has spread and is spreading throughout the world. Moreover ideas of tree breeding were also stimulated (though pursued elsewhere as in Denmark) by, for example, the famous but quite fortuitous cross that occurred in 1905 at Dunkeld in Scotland between the stand of fine European larch and the Japanese larch that had been planted in 1887 by the Duke of Athol to produce the hybrid *Larix eurolepis*.

Summary of progress from 1825 to 1925 in Europe

As we have already seen the basic principles of forest management had been established by 1825 largely owing to the grasp and elucidation of them by G. L. Hartig and later by Heinrich von Cotta. The next hundred years saw a consolidation and refinement of managerial processes under the influence of a better knowledge of the volumes and increment of the growing stock and of the effect of compound interest on length of rotations, as exposed by Faustmann (1849), and a growing conviction of the need for improved silvicultural treatment to suit the condition of individual stands, as exposed by Judeich in 1862.

The silvicultural awareness became more and more marked as the century progressed. This was shown by the writings of Gayer and others, by the number of special silvicultural systems that were designed and by the proposals and action taken to organize the treatment of selection forests after the proposals of Gurnaud. Silvicultural doubts gave rise to study of the growth characters of individual trees, communities of trees and diseases of trees. The knowledge gained and the lines of thought were expressed, disputed and dispersed in the many forestry journals that were founded in most European countries and indeed elsewhere. The development of silvicultural and biological thinking, combined with accidents and failures imposed by unforgiving nature, led to a general disposition to favour more irregularity, mixtures and flexibility in the growing stock and its managerial organization. That tendency was shown in the modification of or departure from periodic block methods, the use of more flexible rotations, more flexible regulation of the yield and less use of pure stands. But on the whole even-aged forestry, though more flexibly applied, with natural regeneration remained supreme.

The influence of transport. It should be remembered that during those hundred years 1825–1925 there was practically no change in the methods of transport and use of power inside the forest. Trees were invariably felled by manpower, by axe and saw. Transport was by cart, drawn by horse or bullock. The only change was some use of tramways and steam-power, some penetration of forests by railways and canals or other waterways. But this penetration of forests by the steam engine has had a smaller lasting influence than the steady extension, intensification and improvement of the road system. The coming impact of the internal combustion engine had not made itself felt by 1925.

Changes in demand and the battle of Hampton Roads. On the other hand changes in the shape of demand for forest produce were becoming noticeable before the end of the nineteenth century. Coal, which had created a huge demand for pit props, had by 1900 in Britain more or less ousted wood fuel; electricity and oil were beginning to compete seriously with coal by 1925. The destruction of the market for wood fuel greatly reduced the value both of the branchwood of hardwood high forest and of coppice which, however, still had some value for fencing and other agricultural uses. But it is often forgotten how serious was the effect of the naval battle of Hampton Roads in 1862. As stated by Albion (1926) the belief that ships of war must be built of iron and not wood had been held by many for some years before 1862. But it was not until the iron *Monitor* had destroyed, by gunfire and explosion of the one and by ramming the other, the wooden ships *Congress* and *Cumberland* which had gallantly resisted the ironclad *Merrimac* the day before, that naval experts finally and unanimously admitted that warships must be built of iron. It had then at last become clear beyond dispute that the gunfire from vulnerable wooden ships was useless against the new armour. Although wooden clippers and other ships of the mercantile marine continued to sail the seas under sail for sometime, the use of hardwood timber as the major constructional raw material for ships was doomed. At the same time coniferous timber was beginning to compete with hardwoods in the other uses which were now found for timber, such as pulp for paper. Concrete and iron were also beginning to compete with strong, desirable hardwoods such as oak for constructional uses.

By 1925, although the total demand for wood continued to increase, it had been recognized in Europe that coppice systems in general were no longer needed. It was beginning to be recognized that coniferous timber would be wanted very much more than hardwoods and, particularly in Britain, that there was a great future for fast growing exotic species such as Douglas fir and Sitka spruce which also yield big volumes. The trend of demand, therefore, in combination with continued development of silvicultural thinking set the stage for wide scale conversion of past methods of silvicultural management. But sustained yields of individual forests and even of parts of a forest such as the working circle and felling series remained an essential element of management.

Summary of developments, 1825 to 1925

The basic developments of European forest management from 1825 to 1925 are listed below:

1. Wide study and application of the means for measuring the volume of trees, stands of trees and their increment exemplified by volume and yield tables, the method of control and classification by quality classes.

2. The introduction, development and use of formulae methods of yield regulation, the study and application of compound interest to management and the emphasis and improvement of methods of keeping records.

3. The continuous study and application of the silviculture of trees, stands of trees, their relations in a forest community and the influence of site factors.

4. The design and application of silvicultural systems which increasingly tended to favour some degree of irregularity, mixtures of species, and more flexible application with a preference for natural regeneration.

5. The conversion of one system to another, particularly coppice with standards to high forest, owing to changes in demand or for silvicultural improvement.

6. Increasing attention to thinning techniques and intensity.

7. The continued general use of indigenous species or their extension to neighbouring localities rather than the introduction of exotic species which were used experimentally only.

8. Sustained yields from individual forests remained a primary element of management.

9. The steady improvement and extension of roads and the trial of new means of transport other than the cart and horse.

DEVELOPMENTS FROM 1925 TO 1967 IN EUROPE AND FUTURE TRENDS

The influence of invention and material advances

To appreciate the progress of forest management since 1925 and to assess future trends it is best to consider first but briefly the explosion of invention, scientific knowledge, material and technological advances that have transformed the way of man's life particularly in those countries that have already been equipped industrially and educationally to use and expand the new developments. The change since 1925 has been so rapid and so vast that it is difficult, to recall and realize how comparatively slow the *tempo* of life then was, how comparatively close were the horizons of most people and how standards of living have altered.

The material and technological progress gained in the last forty years or so have not only been greater but more basic in kind than those gained in the previous 400 years and more. Inevitably the practice of forestry has been greatly affected in the relative values of its products and in the ways and means of producing and delivering them. Yet forestry is still only beginning to absorb and use the new tools and knowledge and to adjust its policies to the new situation. There is also always a danger that in the application of new methods old lessons based on experience and biological truths may be overlooked or underestimated with expensive results.

The main factors of development which have affected and will continue to affect forestry are enumerated in the list below. The various heads are not elaborated, nor is the list exhaustive, nor in any order of priority or suggestive of what factor is likely to influence or cause the next decisive step in forest policy or practice. What, for example, will be the next new use for wood? Will the biologist or bio-chemist or tree-breeder discover means to increase the efficiency of chlorophyll and consequent rates of growth of trees without impairing the properties of wood or exhausting the soil?

Fields of progress which affect forestry and in which development has been marked since 1925

1. Sources of power:
   Oil and nuclear power and thus to electricity with help from coal and water.
2. Mechanical and electrical engineering; electronics.
3. Theory of statistics; mathematical models.
4. Computers and business machines.
5. Chemical theory and products; biochemistry.
6. Plant physiology and ecology.
7. Tree breeding.
8. Timber processing.
9. Integration and enlargement of wood industries.
10. Growth of human populations.
11. Living standards.

Together those fields of social and material progress and advances in biological and other knowledge affect all branches of forestry. For study those branches may be classed under four heads, namely:

(a) Demands for forest produce.
(b) Efficiency of production (rate of producing wood).
(c) Efficiency of supplying produce to consumers.
(d) Efficiency of administration (cost of growing and supplying the produce).

(a) Demands for forest produce

(1) *Material goods.* The trend of world demand for material forest products has already been summarized in chapter II. That summary shows that demands for small round wood (fuel, pit-props, fencing posts and the like) in developed countries are declining, demands for larger sawn-timber are increasing slowly but demands for wood suitable for processing by use of wood particles or fibres, is increasing rapidly.

Consequently coppice in the developed countries of Europe is not wanted (except under special circumstances) and has been under conversion to high forest, often coniferous, for many years particularly since 1925.

It is possible that increasing demands for large saw-timber may not be sustained for long. There has already been competition from substitutes such as iron and even aluminium and even within the wood industry itself laminated beams make it possible to use smaller logs for constructional uses for which conifers have tended to displace hardwoods. Nevertheless new uses and designs for wood in building construction have enabled large saw-timber to hold its own. But it is possible that plastics which are already beginning to displace metals will compete with wood even more seriously than they do and extend their competition to construction load-bearers. On the other hand appreciation of the high decorative value of wood seems likely to continue for a long time.

Therefore it seems probable that the future demands for wood, aided by the outstanding improvements in adhesives, will be for its fibre and mechanically broken particles rather than for its whole wooden structure. Instead of quality, form and size of stem, the quantity of wood produced per acre combined with the quality of fibre such as length and the cell wall—lumen ratio may become the standards of goodness.

Few hardwoods can compete with conifers for the best combination of volume production, regularity of shape and fibre qualities. In the tropics or semi-tropics (frost free) eucalypts can compete, and also *Gmelina arborea* and probably other tropical species. The ability of many conifers in temperate regions to make the best of poor soils and sites is also an advantage to them. Fast growing exotic conifers may also prove to be more profitable (or conversely provide wood fibre more cheaply) than indigenous conifers or tropical hardwoods.

It is worth noting that the forest manager when choosing a species for the production of fibre should not rely only on its MAI by volume. Fibre quality must be considered. Wood density (weight of dry wood) is an excellent indicator with which to assess the combination of quantity and quality of fibre. However, as a measure of comparative value of conifers and hardwoods, density is not advisable owing to the anatomical differences in wood structure. But within conifers the weight of dry wood (e.g. at 12 per cent moisture content) is an excellent criterion and even more so within a given species (e.g. to assess site differences and consequent rates of growth and to assess provenance).

Lavers (1967) gives the weights per cubic foot at 12 per cent moisture content of the wood of many coniferous and hardwood species. Among them are those of four conifers given in Great Britain as shown below.

Sitka spruce	24 lb.
Norway spruce	25 lb.
Douglas fir	31 lb.
Scots pine	32 lb.

The ratio of Scots pine weight to Sitka spruce is $\frac{32}{24} = \frac{4}{3}$. So for the production of wood fibre a maximum MAI of 160 hoppus feet per acre yearly is equivalent to Sitka spruce with a maximum MAI of 180. However Scots pine acquires its maximum in sixty-five years and Sitka spruce in fifty-five years.

The point being made here is that the forest manager in future will increasingly have to study both the quality and quantity production of the species available to him, as well as the time in which the production is attained.

Since 1925, therefore, there has been a definite trend, particularly in Britain where the consumption of coniferous timber is about 90 per cent of the whole, to convert hardwood forests to conifers. Moreover in coniferous forests managed to satisfy the growing fibre market the standard of production will be maximum bulk of useful fibre and not quality of form and structure of stem. On the other hand in coniferous forests grown for saw-timber or special uses quality of the timber as such will remain more important probably than mere volume. The forest manager should therefore make certain what market he is trying to satisfy.

The trends of demand for wood in Europe may be summarized shortly, perhaps over-simply, as below:

(i) Less and less need for hardwood coppice.
(ii) Less need for hardwood saw-timber except that suitable for decorative veneers and furniture.
(iii) Sustained needs for coniferous saw-timber.
(iv) Increasing needs for coniferous wood fibres.

In all countries it is essential to base forecasts of demand on the stage of development that the country has reached and the pace and direction of its progress. In many places in the world the demand for wood fuel and small timber will last for some time yet so that coppice will remain the best silvicultural system for perhaps fifty or more years. There too, as communications and standards of living rise, the need to convert coppice to high forest will come.

From tropical forests the demand for exported high quality decorative timber is likely to continue. If, as is usually true, matched veneers obtained from quarter-sawn logs are wanted, log diameter of some 30 inches and more will be necessary that is to say breast height girths of 8 feet and more. But it may prove uneconomic to grow trees over 9 feet girth.

In tropical forests also the demand for wood fibres will increase and the introduction of fast growing tropical pines and hardwoods such as eucalypts and *Gmelina arborea* has already begun on a substantial scale and will expand. Nevertheless developing countries are likely also to need increasing quantities of constructional saw-timber to meet rising standards of living during a period when plastic or other competitors will enter the market more slowly than in industrially developed countries. For constructional timber it is unlikely that trees larger than 6 feet girth will be needed or be economical to grow. The quality of the timber should be as high as possible.

(2) *Services.* It seems inevitable that the demand for service products of forests will expand steadily.

The service uses of forests have already been considered in chapter II and it is only necessary here to emphasize again the great role that forests must play in providing recreation, quiet and natural beauty for populations that get more and more crowded, more and more confined in an artificial environment, more and more removed from wild fauna and flora which themselves shrink as their habitats disappear.

Higher standards of living and more mechanical, personal transport increase the desires of greater numbers of people to visit forests and open spaces even those at some distance. But in spite of the mobility and ease of transport the forests close to large centres of population will, in particular, have to be managed more and more for amenity. Satisfaction of this demand will undoubtedly exercise foresters to a steadily increasing extent as the years pass. For its satisfaction the growing of commercially valuable stands of even-aged conifers on short rotations may have to be modified. If the sites and soils are good enough, forests set apart primarily for amenity may also be able to grow big and old hardwoods and some gigantic conifers which otherwise could not be grown for the value of their timber alone even if it has decorative, veneer quality.

The trend towards managing forests for amenity, beauty and scientific interest has already begun in many places. Examples are Epping, New and Dean forests in England, the 'Couronne' round Lyon la Forêt in Normandy and preserved tracts of virgin forests in Yugoslavia. The trend will certainly expand in every country in step with industrialization.

(b) Efficiency of production

It is natural that foresters should have always wished and tried to improve the rate of production (i.e. increment) of their forests in both quantity and quality. But now the spur of competition in a world of competing products, investment alternatives, rising populations and stationary areas of land surface injects increased production with paramount importance. The provisos are that increased production must be sustained and permanent and that the means of its attainment must be practical and not inordinately expensive or detrimental to economic harvesting.

Since 1925 two conflicting theories of silviculture have continued to be debated. The one, developing the ideas of the late nineteenth century and supported by ecological study and ably exposed by Köstler (1956) among others, emphasizes the need to consider and treat the forest growing stock as a complex biological community adjusted to its site and in which the continued health of its individual components depend on the harmonious interaction of all; harmony must be maintained if full sustained production is to be achieved. Only a mixed, irregular forest could be the ideal, one composed largely of indigenous species. The other theory is to grow pure even-aged stands artificially raised from species, hybrids or provenances bred to suit the site and objects of management, to improve and adjust soils with fertilizers and, as necessary and possible, to improve sites by manipulation of drainage, sub-soil disturbance and perhaps by irrigation and to combat possible disease with insecticides, fungicides and other means. It may be maintained that with proper silvicultural management and good choice of species disease and failure is not a great danger (Peace, 1960).

One is perhaps unduly influenced by the generally very successful large-scale afforestation, mainly with exotic conifers, that has occurred in Britain since 1925 and is still continuing, by the outstanding work in poplar cultivation in Italy, by the progress in the introduction of tropical pines and eucalypts in the tropics and in frost free areas of Spain. The time that has elapsed since those works began is perhaps still too short for their permanent success to be assured. Nevertheless it seems that the extension of even-aged, artificial forests often of exotic species is certain. One hopes that there will be no repetition of the unfortunate results of the pure spruce plantations of Saxony a hundred years or so ago. Undoubtedly difficulties will arise, perhaps in maintaining soil and site quality, perhaps in combating insect or fungal disease, perhaps in curing un-diagnosed lack of vigour. But scientific knowledge and technological skills are increasing so rapidly that one hopes such difficulties will be overcome without too much expense. It seems inevitable that forestry will copy the artificial progress achieved in agriculture.

That is not to say that the full artificial approach is inevitable or even desirable everywhere. Where indigenous species give a high rate of production and have assured markets their con-tinued cultivation, often in mixture with mutually supporting species, is clearly desirable. Moreover in places where the terrain and concomitant factors such as wind control the silviculture, irregular forestry such as the selection and *femel-schlag* systems using mainly natural regeneration have an assured future. There cleanings and thinnings among the profusion of choice given by natural regeneration and careful selection of the best seed trees will improve the natural strain while artificial cross-pollination might possibly be used to introduce new strains.

Whether fully artificial forestry with even-aged stands or more 'natural' forestry is employed, the forest manager will un-doubtedly try to increase production by more intensive working. Examples are suiting species (and provenance) to site, discourag-ing competitors with arboricides, felling stands at the most appropriate time, thinning stands to concentrate increment on the best stems, filling each unit of land with a full stock of useful trees. Those expedients will, however, tend to fragment stands into smaller areas and thereby make more difficult the economic use of mechanical harvesting. The forest manager will have to choose the right degree of fragmentation and intensity of work after comparing the cost of increased production with the value of the gain.

Means for improving production. The main means by which production has been improved in quantity and quality or cheapened by reduced costs in recent years and are likely to be used more and more in future are summarized below:

1. *Tree breeding* by the discovery and use of hybrids and provenances which give greater volume increment, are more immune from disease or provide higher quality produce.
2. *Introduction of exotics* capable of growing faster than indigenous species and providing products more suitable to the objects of management.
3. *Improvement of soil fertility* by applying fertilizers or other additions in conjunction with physiological knowledge and tests of the species concerned.
4. *Improvement of sites* by drainage, irrigation and penetration of pans. Even substantial adjustment of terrain by huge drag-line machines might be used as exemplified by restoration of agricultural land wrecked by open-cast mining.
5. *Control of insect, fungal and herbaceous pests* by sprays mechanically applied or by biological means.
6. *Control of stocking* by infilling blanks and by concentrating growth on the optimum number of trees so that the maximum increment of the kind desired is obtained at all times.
7. *Removal of stands (or stems)* as soon as they reach the peak of their performance relative to the objects of management.
8. *Mechanization* to cheapen production in nurseries, planting, application of fertilizers and so on.

(c) *Efficiency of harvesting*

The most spectacular changes in forestry since 1925 and particularly after 1946 have been caused by the development of mechanical apparatus and the employment of physical rather than animal sources of power. It is the perfection and varied application of the internal combustion engine that is almost entirely responsible for the changes which include power saws for felling, means of loading produce in the forest and its transport to distant markets. The influence of mechanization is certain to expand.

The more decisive effects of change are:

(i) Removal of the need for sustained yields from individual forests. That has been stressed earlier in this work and is not elaborated again beyond stating that it mitigates one complication of forestry and encourages maximum production (e.g. section (b)7 above).

(ii) Road construction is greatly facilitated.

(iii) The concentration of harvesting is made more desirable. This factor may necessitate a more critical study of the economics of thinnings and also may conflict with intensive silvicultural working and fragmentation of stands. Optimum production (increment) is then sacrificed.

(iv) Interference with surface soil conditions—by compaction and disturbance. This may make subsequent regeneration more difficult and its growth slower.

(v) Damage to surrounding vegetation. This factor may necessitate clear-felling and planting particularly in coniferous forest.

The efficiency of mechanical harvesting is clearly promoted by the clear-felling of large stands, succeeded as necessary by scarifying and loosening the compacted soil, and also by neglect of thinnings. But it is doubtful whether the elimination of thinnings (except early thinnings) would be worth while and it may be hoped that improved mechanical apparatus for felling and converting trees and piling the produce for mechanical extraction without excessive manpower or damage to surrounding trees will be devised. Such devices allied with wide initial spacing of planted trees of selected provenance or clone may provide the solution. More individual care of the fewer trees (e.g. by pruning) will then become necessary but there will certainly be less freedom of choice in thinning to select the desirable stems of the final stand so that the method might only be applicable to stands destined to provide processed wood rather than saw-timber.

The forest manager must always be alert to assess the prospects of applying new methods or apparatus not only to improve and cheapen current methods but to plan his future organization of the forest. Thus if it is found that small pieces of wood can be transported most cheaply by pipe-line, either with water or air as the vehicle of propulsion, he may find it necessary to segregate those parts of the forests devoted to supplying wood for processing. Moreover the cheaper cost of transport may be counterbalanced by the extra cost of converting logs in the forest to transportable size, such as chips; large clear-fellings might be the only justification for which long-term planning must be prepared.

(d) Efficiency of administration

Besides those developments which have been affecting demands for forest products, the production of forest produce and its harvesting and delivery there have been many others which have influenced forestry as a whole since 1925 and will continue to have a strong influence.

Perhaps the most outstanding new influence on forestry has been statistical analysis whose rapid development resulted largely from the work of Fisher (1925) on statistical methods. In consequence not only have research methods and the study of research results been profoundly affected but the whole business of measuring the growing stock of a forest has been transformed. More recently but complimentary to that basic advance the electronic computer has enabled calculations to be made and compared with incredible speed and the results stored for future use. The combination of statistical analysis and computer ability to handle and interpret data rapidly and accurately has made it possible by sampling with a calculable degree of precision to collect more data of all kinds more cheaply, analyse them more thoroughly and compare many alternative courses of action more easily and certainly. As the task of a manager is to examine facts, assess them and then make decisions, those advances in theory and technology should greatly enhance his efficiency. He has to decide what facts are to be measured, how those facts affect each other quantitatively, prepare the consequent mathematical model and formulae and abstract the answer from a computer. Even that over-simplification of the series of events shows how complex the matter can be.

Co-incident with and linked to the advances of mechanization has been the introduction of work and method studies in operational research. Formal work studies in forestry works were begun in Sweden in the 1930s and in Britain in 1956. Research in this field has far-reaching effects on efficiency and its expansion seems certain. On it depends not only the proper use of tools but mechanical adaptation of the tools and design of new tools and methods.

The introduction of aerial photography and the great advances made in the science of photogrammetry and photo interpretation, particularly after the 1939–45 war has introduced a new aid to mapping and assessing forest growing stocks. Although aerial photographs themselves may not yet be able to give, particularly in hardwood, tropical forests, sufficiently detailed and accurate information of species and timber volumes for any but very extensive management, it is evident that their value for mapping forest types and assessing the broad potential of a large, remote and inaccessible forest is great. But even in smaller, more accessible forests they can hasten and cheapen inventories by providing at least a preliminary stratification on which a ground sampling project can be organized. They also provide means for a quick assessment of damage such as that from wind, fire, insects, fungi or drought. Without elaborating the value of aerial photographs, whose use is still under active development, it is clear that they are a new and valuable tool to the forest manager.[1]

The advent of a whole range of business machines has also presented to the manager new aids for expediting and facilitating work, for assessing, distributing and recording instructions and information. The difficulty which faces the manager (of whatever magnitude) is for him to have the ability and time to grasp the mass of detail that can be made available and to understand and distinguish the degree of interdependence and individual importance of a complex of component factors. The human factor of a man being able to cope with a mass of varied activities, assess their relative values and make sound decisions remains and will always remain the prime influence on success or failure.

The internal organization of an enterprise must, therefore, be such that managers at each stage of its structure have full

---

[1] Even in small, intensively worked artificial forests aerial photographs may reveal what has been unnoticed on the ground. In aerial photographs of the New Forest in England the writer noticed differences in the depiction of a stand of twenty-year-old Corsican pine. Reinspection on the ground showed that the trees on either side of drainage ditches were markedly taller and stouter than those between the ditches.

opportunities to make sound decisions. Modern developments are making forestry more complex so that the organization of forest enterprises are being critically studied and changes made.

Four aspects in the organization of a forest enterprise seem critical namely:

1. Increased use of specialists.
2. Improved accountancy, linked with
3. Improved collection of data and keeping of records.
4. Integration of the management of small forests.

As the author does not feel sufficiently competent to present those aspects in any detail and there is not in any event sufficient space, only a few comments on them will be made.

1. *Increased use of specialists.* The forest manager, particularly the managing director, cannot be expert in all branches of forestry although he should have a wide grasp and experience of its several components. He must, therefore, depend on specialists and experts who will study and collect information and work in their own fields and present sifted, considered appreciations to their director to enable him to distinguish the more important from the less important among conflicting evidence and requirements of the different branches and to decide on future courses of action. Moreover the individual experts by conducting work in their own field of knowledge and experience increase the efficiency of it. The director co-ordinates the work of all.

The Forestry Commission of Great Britain has recently (1965) been re-organized on those lines so that the Commission itself and each regional Conservancy includes these specialist branches on Administration and Finance, Harvesting and Marketing, Forest and Estate Management.

It may be that that organization does not give enough importance to the technical and biological side of growing trees, i.e. to silviculture in its widest sense to include all factors affecting the growth and tending of trees from sites to diseases. To overcome that deficiency a fourth branch, Silviculture, could be added to the existing three branches. Alternatively the branch of Forest and Estate Management might be called Forest Operations and be expanded to include sub-branches as shown in Table 31.

In any such division of responsibilities the main difficulty is to retain the essential chain of command to ensure co-ordination of action between branches without interfering with initiative within a branch. The great function of the managing director is to achieve that co-ordination and promote initiative and enthusiasm and to be able and willing to choose and decide the essential and critical priorities without interference in lesser matters. To enable him to do that the organization must suit the functions of the enterprise and their relative importance and have good communications.

Table 31

ADMINISTRATIVE ORGANIZATION AT THE TOP

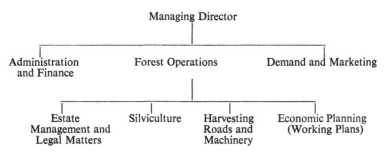

	Managing Director	
Administration and Finance	Forest Operations	Demand and Marketing
Estate Management and Legal Matters	Silviculture	Harvesting Roads and Machinery	Economic Planning (Working Plans)

2. *Improved accountancy.* There are perhaps three basic purposes of a system, namely that it should display:

(i) The capital structure of the forest enterprise.
(ii) The network of expenditure.
(iii) The sources of income.

The peculiarities of and difficulties in assessing the value of the growing stock and its changes have already been considered in chapter IX and will not be elaborated any more.

In records of expenditure and income it is clearly desirable for the accounts to expose clearly the directions in which the money goes and comes so that efforts can be made by cost analyses and work and method studies to reduce costs and improving harvesting and sales to increase income. But whenever possible it is also desirable to link expenditure with income and show how and to what extent increased expenditure increases incomes. Will better drainage, for example, increase increment and consequent income in proportion to the cost of improved drains?

It is again the time factor in forestry that makes it difficult to discover the answers such as whether better drains are economically justified. Special research into the economic effect of drains can, of course, be initiated once the accounts have established that drains constitute a substantial element of costs worthy of detailed study. But the problem also suggests that accounts should be linked closely with the progress of the growing stock established from records maintained by compartments, working blocks or, best of all, by site or forest types. Accounts are usually classified by particular activities such as drainage, cleaning or pruning and then consolidated for an administrative unit such as a forester's charge. But a sub-classification of activities by site or forest types would clearly be advantageous although the increased complication and cost of the increased detail of accountancy are obvious.

What is clear is that the accountancy network should be devised after very close consultation between a forester and an accountant so that it fits the administrative organization of a forest, the manner of growing and harvesting the produce and the system of recording changes in the growing stock.

3. *Improved record keeping.* The primary objects of records of the growing stock are to show:

(i) What is there—quantity, quality and, if possible, value by age or other classes as necessary.

(ii) How it is performing—rate of increment in quantity and value.

(iii) What is harvested—and its value.

(iv) The treatment accorded—and cost.

(v) The effect of treatment on the performance of the stock.

Histories or registers of compartments subjected to repeated inventories and annual record of activities are capable of providing the information of items (i), (ii) and (iii) without great difficulty. To complete item (iv) is more difficult as supervision of work and costs and their classification by small spatial units may be time consuming and expensive. To satisfy item (v) is more difficult as a compartment may contain several site, specific or forest and stand types. To lump all types together and apply results of treatment to all equally is clearly misleading.

One is led to the conclusion that the basis of records should be the site-type sub-classified if really necessary by specific or forest types. Only then can records of treatment be related adequately to the performance of the growing stock in various conditions. Compartments should then be large or abolished and replaced by site-types. One can visualize an intensively managed forest that contains, say fifteen carefully distinguished site-types labelled alphabetically A to O. Individual site-types would then be numbered serially according to locality (A1, A2, etc.).

Records of treatment and costs relative to various types of growing stock should then be more easily handled. But re-organization of any forest whose management is based on compartments into one managed by site-types would be a big undertaking and also requires very detailed study and classification of soils and sites; it is probably only worth while in very intensively managed forests or in forests that have pronounced site differences. Large compartments or parcels in which site types are distinguished is an alternative (see p. 118). Alternatively it may be found that the best information on the effect of treatment on performance of stock is obtainable only by detailed research study. Nevertheless continual records of actual costs of growing forest produce in different site-types would always be very useful.

4. *Integration of forest units.* In chapters III and VIII and elsewhere it has been shown that in industrially developed countries sustained yields are not essential from individual forests but only from a region containing perhaps six separate forests. Together the six forests can deliver regular supplies to the consumers of forest produce although an individual forest may become responsible for supplies of a particular forest produce for which it is best suited.

The management of several forests can then be integrated in a way similar to the current trend of amalgamating comparatively small industrial firms in a combine managed by a large holding company. Such an integration of forest management should facilitate harvesting and marketing especially for supplies to large wood-processing industries when new trends in the shape, quality and quantity of wood as a raw material are developing; it should also facilitate the use of large and expensive equipment and give greater freedom to the silviculturist to attain better growth as he would not be tied so closely by the constraint of a sustained yield from a small forest unit. There should also be a reduction of overhead costs so that efficiency in general is promoted provided that supervision everywhere remains adequate.

The combination of several forests into an integrated forest region would require a regional working plan in which the individual forests would have subsidiary plans similar to those for working circles or felling series of the classic working plan. But perhaps a greater consequence of integration would be the administrative changes. To obtain full advantage of centralization it seems likely that administrative charges could also be enlarged.

Larger managerial charges (e.g. of the professionally trained district officer) and the consequent wider responsibilities entailed would result in the manager having less time to know his forest and supervise work in the field. Yet the technical aspects of forestry are becoming more specialized and intensive. Consequently there would be a greater need not only for specialists for advice and research but also for highly qualified and able sub-managers (foresters or rangers) who have been trained to a higher standard than in the past and are supported by able, trained assistants so that they can execute and supervise intensive silviculture, harvesting and estate management.

Although integration of management is feasible for State forests it is much more difficult, but perhaps even more desirable, for private owners whose many forests may also be composed of smaller woods and be more scattered over a region than the State forests. Yet some degree of integration would almost certainly improve commercial efficiency and the income of the owners. The directions in which co-operation between owners can expand are by means of marketing associations and consultant and executive forestry companies or societies such as those that have been developing in Britain. Some forms of co-operation (if only by exchange of information on markets and forecasted yields) between private owners and the State also seems desirable; combined private and State sale depots is a possibility.

In summary the trend must be towards making the practice of forestry more efficient and the commercial trend of large units must have its counterpart in forestry without which supplies of wood to the large wood industries must be more complicated and inefficient. That is not to say that there is no longer any scope in industrially developed countries for the efficiently managed small forest. Where small forests, particularly private or communal forests, satisfy keen local demands, particularly for a specialized product on which local welfare depends or where other larger forests are fully occupied in supplying standard products for big industry, there may be substantial opportunities for the comparatively small, independent forest provided that it is not too small and independent to bear the costs of skilled supervision and management.

Conclusion

That account of the historical development of forestry has been confined almost entirely to European forestry since it shows the stages of progress from early empirical and comparatively extensive methods to the intensive cultivation and complicated management which is developing there today. The progress has marched in step with general social standards and organization, with changes in silvicultural and biological knowledge, in industrial development, demands and mechanical inventions. The lesson to be learnt is that the management of any forest must be influenced by its past and particularly its immediate past and consequent present structure, composition and accessibility; it must suit the silvicultural and other knowledge, skills and resources available; it must anticipate the prospects of future demands on its products.

It is probable that certain stages experienced in Europe will be omitted elsewhere and that silvicultural and other methods will differ substantially in detail. Nevertheless the progressive phases of European forestry should provide lessons useful to all.

It seems likely, as stated earlier, that tree-breeding, the introduction of exotics, particularly conifers, artificial regeneration, mechanization, chemical aids and site improvement will be used everywhere to a greater and greater extent with more and more reliance on the soil scientist, plant breeder, plant physiologist and plant pathologist to maintain and improve tree health and vigour. In forest administration, forest charges, initially very large and inaccessible and extensively managed, will be broken into smaller and perhaps again yet smaller and more intensively managed forests that have been made fully accessible with a network of roads and rides. There may then come an integration of several forests into a regional consortium of forests managed under a general plan with subsidiary plans for its component forests. Such integration will be made as and when the general industrial development of the country and the shape of demand for wood makes it necessary, as is occurring in Britain today.

Bibliography

Albion, R. G. (1926) *Forests and Sea Power*. Harvard University Press, Cambridge, Mass.

Dwight, T. W., Jaciev, P. (1965) Confusion Worse Confounded. The Sad Story of Forest Regulation. *The Forestry Chronicle*, (41)

Faustmann, M. (1849) Berechnung des Werthes welchen Waldboden sowie noch nicht haubare Holzbestande für die Waldwirtschaft besitzen'. *Allgemeine Forst und Jagd-Zeitung*, 15. (pp. 411–455)

Fisher, R. A. (1925) *Statistical Methods for Research Workers*. Oliver and Boyd, Edinburgh

Huffel, G. (1926) *Économie Forestière*, Tome III. Librairie de l'Académie d'Agriculture, Paris

Köstler, J. (1956) *Silviculture*, translated by Mark L. Anderson. First English edition, Oliver and Boyd, Edinburgh

Lavers, G. M. (1967) 'The Strength Properties of Timbers', *Ministry of Technology*, *Forest Products Research, Bulletin No 50*. HMSO, London

Maury, L. (1850) *Histoire des grandes forêts de la Gaulle et de l'ancienne France*. A. Leleux, Paris

Peace, T. R. (1960) 'The Dangerous Concept of the Natural Forest', an address to the British Association of Science. Republished (1961), *The Quarterly Journal of Forestry*, IV (1)

Pressler, M. R. (1868) *Zur Forstzuwachskunde*. Woldemar Tunks Verlagshandlung, Dresden

APPENDIX I

(Extracts from *Working Plans Code*, reproduced with
permission)

(*a*) Sub-compartment form

(*b*) Compartment record

(*c*) Area of growing stock

(*d*) Area analysis by yield class

(*e*) Forecast of allowable cut—thinning and felling

(*f*) Production programme and control

(*g*) Forecast of allowable cut

FORESTRY COMMISSION

Form WP.O
(Revised 12/64)

Appendix I (a)

SUB-COMPARTMENT ASSESSMENT FORM

W.P. Area .... Compt. .... Acreage: ....

Section or Beat: .... Sub Compt. .... Acreage: ....

LAND USE	H.F.			L.F.N.P.							
	Sat.	Unsat.	Felled	Cop.	Scrub	Bare	Reserved	Nurs.	F.W.H.	Agric.	Unusable	Other

AGE CLASS STRUCTURE	EVEN	UNEVEN		
		Post 1900	Pre.1900	All

DESCRIPTION	Stocking %	Species	Species %	% in check	P.yr.	Adj. P.yr.	Top Ht.	G.Y.C.	Mean Girth Top Ht. Trees	L.Y.C.
MAIN STOREY										
OVER STOREY				/////		/////				

REMARKS: .... ------ Surveyor

Appendix I (b)

Form WP.I
(Revised 12/64)

FORESTRY COMMISSION

COMPARTMENT RECORD

Location.....................................   Cpt. No..............

Crop Assessment as at................ 19..........   Area (Acres)..............

SUB. CPT.	AREA (ACRES)	SPECIES / LAND USE	IDENTITY NUMBER	P. YEAR	GENERAL YIELD CLASS	LOCAL YIELD CLASS
(1)	(2)	(3)	(4)	(5)	(6)	(7)

General Notes
(inc. unusual site factors, previous land use, establishment procedure, etc.)

A. Previous Treatment
Information from previous records which is still relevant

B. Treatment During Current Plan Period
Year thinned, intended cycle, fertiliser prescription.
Details of any unusual treatment.

Appendix I (c)

FORESTRY COMMISSION

AREA STATEMENT OF GROWING STOCK AT AT 30.9.19

Working Plan Area.................................

Complete as necessary—Section or Beat.................................

Delete as necessary { High Forest—Main Storey / High Forest—Over Storey / Coppice

Areas in Acres

Species (1)	EVEN-AGED			AGED		UNEVEN-AGED			Total (20)
	1966 1970 (2)	1961 1965 (3)	1956 1960 (4)	1841 1860 (15)	Pre 1841 (16)	Pre 1900 (17)	Post 1900 (18)	† Pre and Post 1900 (19)	

†Uneven-aged crops in which pre P.1900 and post P.1900 age groups each occupy at least 20% of the canopy.

Signature.................................

(Survey Team Leader)

Date of Compilation.................................

FORESTRY COMMISSION

Form WP.6 (Revised 12/64)

Working Plan Area: *Example*

AREA ANALYSIS BY YIELD CLASS AS AT 30.9.19 64

SPECIES	P.YEAR CLASS	300	280	260	240	220	200	180	160	140	120	100	80	60	40	20	CHECK	TOTAL AREA (AC.)	WEIGHTED AV. YIELD CLASS	REMARKS
(1)	(2)	(3)	(4)	(5)	(6)	(7)	(8)	(9)	(10)	(11)	(12)	(13)	(14)	(15)	(16)	(17)	(18)	(19)	(20)	(21)
SS	61-70							355										355	180	
	51-60							1047										1047	180	
	41-50						1	111	550	92							85	839	160	
	31-40						319	1985	276	14							156	2750	180	
	21-30					14	286	863	269	31								1463	180	
	11-20				7	43	102	51										203	200	
	01-10					13	49	6										68	200	
	Pre 1900																			
	Totals				7	70	757	4418	1095	137							241	6725	180	
							OVERSTOREY CROPS													
	61-70																			
	51-60																			
	41-50																			
	31-40									25	102	36						163	120	
	21-30									8	53	13						74	120	
	11-20																			
	01-10																			
	Pre 1900																			

Signature: *J Smith* (Survey Team Leader)

Date of Compilation: 3/64

Appendix I(e)

FORESTRY COMMISSION

Form WP.17
(Revised 12/64)

FORECAST OF ALLOWABLE CUT — THINNING FELLING

Directorate/Conservancy
(delete as appropriate)

Working Plan Area

Section

Date of
Compilation

VOLUMES IN 000'S H.FT. OVER BARK

VOLUME CATEGORY	YEAR OF FORECAST	Grand Total (all Species)	CONIFEROUS				BROADLEAVED			
			Total Coniferous	Scots Pine	Douglas Fir	Other Conifers	Total Broadleaved	Oaks	Beech	Other Broadleaved
(1)	(2)	(3)	(4)	(5)	(12)	(13)	(14)	(15)	(16)	(17)
VOLUME TO 3" TOP DIAM. OVER BARK										
VOLUME TO 7" TOP DIAM. OVER BARK										
VOLUME TO 8" TOP DIAM. OVER BARK										

Species and Species Groups

* min. log length 10ft.

Signature .... (District Officer)   Date of Compilation ....

Appendix 1(f)

PRODUCTION PROGRAMME AND CONTROL

THINNING
FELLING

F.Y. _____

Working Plan Area _____

Volumes in 000's h.ft. over bark except in cols. marked with •

Compt. or Sub.Cpt.	Species	Age	Yield Class	Annual Thinning *Yield/Acre	Thinning Cycle	Volume per Acre	Area	Over 3"	Over 7"	Over 9"	Area	Vol.	Area	Vol.	Vol.	Standing Sales		
								LOCATION AND VOLUME				Marked		Completed in F.Y.		CONTROL	PROPOSED DISPOSAL		
								Volume by Diameter Classes											
1	2	3	4	5*	6	7*	8	9	10	11	12	13	14	15		16	17	18
Totals																		

Allowable Cut: _____

Signature _____

Date of Compilation _____

Appendix *I(g)*

WORKING PLAN AREA: **Example**

PART A — AREAS of various categories by year

P. Year Class	Present Areas (ac.) Normal	Over Storey	1965 Check	Non-thin	Thin-ning	Over Storey	Felled	1970 Check	Non-thin	Thin-ning	Over Storey	Felled
(1)	(2)	(3)	(4)	(5)	(6)	(7)	(8)	(9)	(10)	(11)	(12)	(13)
61–70	355			355					355			
51–60	1047			1047					1047			
41–50	839		85	528	226			55	157	627		
31–40	2750	163	156		2594	163				2750	162	1
21–30	1463	74			1463	74				1463		74
11–20	203				203				30	91		82
01–10	68			30	38							68
Pre-1900												
TOTALS	6725	237	241	1960	4524	237		55	1589	4931	162	225

ANNUAL ACREAGE OF U/Planting 1. and Felling 2. — 1.   2. **30**   1.   2. **30**

ANNUAL ACREAGE OF Overstorey Fellings — 3.   4. **15**   3.   4. **15**

PART B — Conversion of areas

P. Year Class	Wtd. Av. Yield Class	Av. Age	Areas	Vol 3"	Vol 7"	Vol 9"	Av. Age	Areas	Vol 3"	Vol 7"	Vol 9"
(29)	(30)	(31)	(32)	(33)	(34)	(35)	(36)	(37)	(38)	(39)	(40)
THINNING YIELD											
61–70	180 (21)										
51–60	180 (21)										
41–50	160 (22)	23	226	22	—	—	25	627	60	—	—
31–40	180	30	2594	280	29	—	35	2750	297	74	11
21–30	180 (55)	40	1463	158	72	19	45	1463	158	99	41
11–20	200 (55)	49	203	24	20	13	53	91	11	10	7
01–10	200 (55)	57	38	5	4	3					
Pre-1900											
THINNING TOTAL				489	125	35			526	183	59
1. UNDER-PLANTING YIELD											
2. FELLING YIELD		65	30	171	164	152	55	30	152	142	124
3/4. O/STOREY YIELD		45	15	42	38	29	40	15	38	30	18
FELLING TOTAL				213	202	181			190	172	142
TOTAL ALLOWABLE CUT				702	327	216			716	355	201

Appendix I(g)

ALLOWABLE CUT

FORM WP 20 (Rev'd 12/64)

Species: _Sitka Spruce_

of forecast (Even-aged High Forest Plus Overstorey of Two-storey Forest)

1975					1980					1985				
Check	Non-thin	Thinning	Over Storey	Felled	Check	Non-thin	Thinning	Over Storey	Felled	Check	Non-thin	Thinning	Over Storey	Felled
(14)	(15)	(16)	(17)	(18)	(19)	(20)	(21)	(22)	(23)	(24)	(25)	(26)	(27)	(28)
	355					355					213	142		
	628	419			105	942						1047		
		839					839					789	50	
		2750	87	76			2750	12	151			2750		163
100		1334		103	100		834		603	200		234		1103
				203					203					203
				68					68					68
	1083	5342	87	450		560	5365	12	1025		413	4962	50	1537

Footnote boxes:
- 1975: 1. — 2. 100 3. — 4. 15
- 1980: 1. 10 2. 100 3. — 4. 12
- 1985: 1. 10 2. 200 3. — 4. —

into VOLUME YIELDS by year of forecast (000's H.ft.)

Av. Age	Areas	Volume by Top Diams.			Av. Age	Areas	Volume by Top Diams.			Av. Age	Areas	Volume by Top Diams.		
		3"	7"	9"			3"	7"	9"			3"	7"	9"
(41)	(42)	(43)	(44)	(45)	(46)	(47)	(48)	(49)	(50)	(51)	(52)	(53)	(54)	(55)
										21	142	15	—	—
22	419	45	—	—	25	942	102	2	—	30	1047	113	12	—
30	839	81	4	—	35	839	81	13	1	40	789	76	25	4
40	2750	297	135	36	45	2750	297	187	77	50	2750	297	226	124
50	1334	144	109	60	53	834	90	73	45	55	234	24	21	14
		567	248	96			570	275	123			525	284	142
					40	10	12	7	2	45	10	15	11	6
55	100	463	424	347	55	100	463	424	347	55	200	926	848	694
40	15	38	30	18	40	12	30	24	14					
		501	454	365			505	455	363			941	859	700
		1068	702	461			1075	730	486			1466	1143	842

APPENDIX II

EXTRACTS FROM BRITISH FOREST MANAGEMENT TABLES

From Forestry Commission Booklet No. 16 by R. T. Bradley, J. M. Christie and D. R. Johnston, published by HMSO, London, with permission

(a) Thinning control tables:

Sitka spruce. Yield classes 100 to 180

(b) Production forecast tables:

(i) Thinning yields.
Sitka spruce. Yield classes 140, 160

(ii) Felling yields.
Sitka spruce. Yield classes 140, 160

(c) Normal yield tables:

(i) Sitka spruce. Yield classes 160 to 200

(ii) Douglas fir. Yield classes 160 to 200

Appendix II (a)

SS 180   THINNING CONTROL TABLE   **SS 100**

Sitka Spruce

Age	Yield Class 180			Yield Class 160			Yield Class 140			Yield Class 120			Yield Class 100			Age
	Vol. per acre per annum, h. ft.	Basal area per ac. per ann. sq. ft. q.g.	Tariff Number	Vol. per acre per annum, h. ft.	Basal area per ac. per ann. sq. ft. q.g.	Tariff Number	Vol. per acre per annum, h. ft.	Basal area per ac. per ann. sq. ft. q.g.	Tariff Number	Vol. per acre per annum, h. ft.	Basal area per ac. per ann. sq. ft. q.g.	Tariff Number	Vol. per acre per annum, h. ft.	Basal area per ac. per ann. sq. ft. q.g.	Tariff Number	
18	67	6.9	16													18
19	92	8.7	17	59	6.4	16										19
20	111	9.6		82	8.1		52	6.0	15							20
21	126	10.4	18	99	9.1	17	72	7.6	16							21
22	126	9.6	19	112	9.9	18	86	8.5		45	5.3	15				22
23	126	8.9	20	112	9.1	19	98	9.3	17	61	6.6	16				23
24	126	8.3	21	112	8.4		98	8.5	18	74	7.5		37	4.6		24
25	126	7.8	22	112	7.8	20	98	7.9	19	84	8.1	17	51	5.9	15	25
26	126	7.4		112	7.4	21	98	7.4		84	7.5		62	6.5		26
27	126	7.0	23	112	7.0	22	98	7.0	20	84	7.0	18	70	7.1	16	27
28	126	6.7	24	112	6.6		98	6.6	21	84	6.6	19	70	6.6	17	28
29	126	6.4	25	112	6.3	23	98	6.3		84	6.2		70	6.2	18	29
30	126	6.1		112	6.0	24	98	6.0	22	84	5.9	20	70	5.8		30
31	126	5.9	26	112	5.8		98	5.7	23	84	5.6	21	70	5.5	19	31
32	126	5.7	27	112	5.6	25	98	5.5		84	5.4		70	5.3		32
33	126	5.4		112	5.3	26	98	5.2	24	84	5.1	22	70	5.0	20	33
34	126	5.2	28	112	5.1		98	5.0		84	4.9		70	4.8		34
35	126	5.1	29	112	5.0	27	98	4.9	25	84	4.8	23	70	4.7	21	35
36	126	5.0		112	4.8		98	4.7		84	4.6		70	4.5		36
37	126	4.8	30	112	4.7	28	98	4.6	26	84	4.5	24	70	4.4	22	37
38	126	4.6		112	4.5	29	98	4.4	27	84	4.3		70	4.2		38
39	126	4.5	31	112	4.4		98	4.3		84	4.2	25	70	4.1	23	39
40	126	4.4	32	112	4.3	30	98	4.2	28	84	4.1		70	4.0		40
41	126	4.3		112	4.2		98	4.1		84	4.0	26	70	3.9	24	41
42	126	4.2	33	112	4.1	31	98	4.0	29	84	3.9		70	3.8		42
43	126	4.1		112	4.0		98	3.9		84	3.8	27	70	3.7	25	43
44	126	4.0	34	112	3.9	32	98	3.8	30	84	3.7		70	3.6		44
45	126	3.9		112	3.8		98	3.7		84	3.6	28	70	3.5		45
46				112	3.7		98	3.6		84	3.5		70	3.4	26	46
47			35	112	3.6	33	98	3.5	31	84	3.4	29	70	3.3		47
48							98	3.4		84	3.3		70	3.2		48
49			36			34			32	84	3.2		70	3.1	27	49
50	110	3.2	37	102	3.2		93	3.1		84	3.2	30	70	3.0		50
51													70	2.9	28	51
52						35										52
53									33			31				53
54			38												29	54
55	95	2.6		87	2.5	36	79	2.5	34	70	2.4		61	2.3		55
56																56
57			39									32				57
58						37			35						30	58
59												9				59
60	83	2.1		75	2.0		68	2.0		60	2.0	33	52	1.9		60
61			40			38										61
62									36						31	62
63																63
64												34				64
65	73	1.7	41	67	1.7	39	60	1.7		53	1.7		46	1.6		65
66									37							66
67															32	67
68																68
69												35				69
70				59	1.5	40	54	1.5	38	47	1.4		41	1.4		70
71															33	71
72																72
73																73
74																74
75										43	1.3	36	39	1.3	34	75

Appendix II (b) (i)

SS 160   PRODUCTION FORECAST TABLE   **SS 140**

Sitka Spruce

THINNING YIELDS

Age	Mean BHQG ins.	Yield Class 140 Volume (h. ft.) to top diameter o. b. of 3 inches	7 inches	9 inches	Mean BHQG ins.	Yield Class 140 Volume (h. ft.) to top diameter o. b. of 3 inches	7 inches	9 inches	Age
21									21
22	4	96							22
23		96			4	84			23
24		96				84			24
25		96				84			25
26		96	1			84			26
27		96	2			84			27
28		96	3			84	1		28
29		96	4			84	1		29
30		96	5			84	2		30
31		96	7			84	3		31
32		96	8			84	3		32
33		96	10			84	5		33
34	5	96	13	1		84	6		34
35		96	15	1		84	7		35
36		96	18	1		84	9		36
37		96	21	2	5	84	11		37
38		96	25	3		84	13	1	38
39		96	28	4		84	15	1	39
40		96	32	5		84	17	1	40
41	6	96	35	7		84	19	2	41
42		96	38	8		84	22	3	42
43		96	41	10		84	25	3	43
44		96	45	12		84	27	4	44
45		96	48	14	6	84	30	5	45
46		96	52	16		84	32	7	46
47		96	55	18		84	35	8	47
48	7	96	57	20		84	37	9	48
49		96	60	22		84	40	11	49
50		96	62	25		84	42	12	50
51		96	64	27		84	45	14	51
52		96	66	30		84	47	16	52
53		96	68	32		84	49	18	53
54		96	70	35	7	84	51	19	54
55	8	96	72	38		84	53	21	55
56		96	73	40		84	55	23	56
57		96	75	43		84	56	24	57
58						84	58	26	58
59						84	59	28	59
60									60
61									61
62									62
63									63

Appendix II (b) (ii)

SS 160 PRODUCTION FORECAST TABLE **SS 140**

Sitka Spruce

FELLING YIELDS

Age	Yield Class 160				Yield Class 140				Age
	Mean BHQG ins.	Volume (h. ft.) to top diameter o. b. of 3 inches	7 inches	9 inches	Mean BHQG ins.	Volume (h. ft.) to top diameter o. b. of 3 inches	7 inches	9 inches	
25	4	1300	50						25
30	5	1800	300		4½	1550	150		30
35	6	2350	900	200	5½	2050	500	50	35
40	7	2900	1800	700	6½	2550	1150	300	40
45	8	3400	2600	1350	7¼	3000	1900	750	45
50	8¾	3850	3200	2050	8	3400	2600	1350	50
55	9½	4200	3700	2700	8¾	3750	3100	1950	55
60	**10¼**	**4500**	**4100**	3250	9¼	**4050**	**3500**	**2450**	60
65	10¾	4800	4400	3700	9¾	4300	3800	2900	65
70	11¼	5050	4700	4050	10¼	4550	4100	3250	70
75	11¾	5250	4950	4350	10½	4750	4350	3600	75
80	12	5400	5100	4600	10¾	4900	4500	3850	80

Appendix II (c) (i)

SITKA SPRUCE

NORMAL YIELD TABLE: YIELD CLASS **200**

Age	MAIN CROP After Thinning							Yield From THINNINGS						TOTAL Production		INCREMENT			Age
	Number of Trees	Height Top feet	Mean BHQG ins.	Basal Area sq. ft. q. g.	Volume (h. ft.) to top diameter o.b. of 3 inches	7 inches	9 inches	Number of Trees	Mean BHQG ins.	Av. Vol. per Tree h. ft.	Volume (h. ft.) to top diameter o.b. of 3 inches	7 inches	9 inches	Basal Area sq. ft. q. g.	Volume to 3 inches h. ft.	C.A.I. Basal area	C.A.I. Volume to 3 inches	M.A.I. Volume to 3 inches	
																	5		
15	1300	25	3¼	91	825	—	—	—	—	—	—	—	—	91	825	10.0	165	55	15
20	825	35½	3¾	86	1110	20	—	475	4	1.46	700	15	—	138	1810	8.9	233	91	20
25	530	46	4¾	89	1715	210	10	295	4¼	2.37	700	40		181	3115	7.9	281	125	25
30	376	56½	6	98	2485	960	200	154	5	4.54	700	115	10	218	4585	6.9	299	153	30
35	284	66½	7¼	106	3280	2180	890	92	6	7.65	700	275	55	250	6080	5.9	295	174	35
40	26	75	8½	113	4015	3250	1960	58	7	12.2	700	430	150	278	7515	5.0	277	188	40
45	189	82½	9½	120	4625	4070	3000	37	8	18.7	700	530	285	301	8825	4.2	247	196	45
50	164	89	10½	126	5195	4750	3910	25	9	23.2	590	500	340	320	9985	3.6	217	200	50
55	147	94½	11¼	131	5700	5320	4650	17	9¾	29.1	510	450	355	337	11000	3.1	188	200	55
60	134	99	12	136	6135	5820	5250	13	10½	35.1	445	410	340	352	11880	2.7	163	198	60
65	124	103	12¾	140	6505	6220	5730	10	11¼	41.1	390	375	320	365	12640	2.3	142	194	65
70	116	106½	13¼	143	6815	6570	6140	8	12	46.6	350	330	295	376	13300	2.0	124	190	70
75	110	109½	13¾	146	7085	6850	6450	6	12½	52.4	310	295	270	385	13880	1.7	109	185	75
80	105	112	14¼	148	7320	7100	6700	5	13	57.5	275	265	250	393	14390	1.5	94	180	80

YIELD CLASS 180

15	1300	22½		79	645	—	—	—	—	—	—	—	—	79	645	9.5	130	43	15
20	960	32	3½	87	970	—	—	340	4	1.47	500	10	—	125	1470	8.6	199	73	—
25	635	42	4½	89	1465	100	—	325	4¼	1.94	630	25	—	166	2595	7.7	246	104	25
30	442	52	5½	95	2135	540	70	193	4¾	3.27	630	60	—	202	3895	6.8	270	130	30
35	334	61	6¾	103	2880	1520	440	108	5½	5.86	630	170	20	234	5270	5.8	274	150	35
40	264	70	7¾	110	3595	2600	1250	70	6½	9.06	630	300	75	261	6615	4.9	260	165	40
45	219	77	8¾	117	4195	3480	2230	45	7¼	13.8	630	410	165	284	7845	4.1	232	174	45
50	188	83	9¾	122	4740	4190	3120	31	8	18.0	550	425	235	303	8940	3.5	205	179	50
55	167	88½	10½	127	5225	4770	3890	21	8¾	22.7	475	400	265	319	9900	3.0	178	180	55
60	152	93	11¼	131	5640	5250	4530	15	9½	27.5	415	365	270	333	10730	2.6	154	179	60
65	141	97	11¾	135	6000	5640	5040	11	10¼	32.0	365	330	265	345	11455	2.2	134	176	65
70	132	100	12¼	138	6295	5990	5430	9	10¾	36.7	330	300	250	356	12080	1.9	117	172	70
75	125	103	12¾	141	6555	6280	5760	7	11¼	40.9	290	270	235	365	12630	1.6	103	168	75
80	119	105½	13¼	143	6765	6500	6040	6	11¾	45.0	265	245	220	373	13105	1.4	90	164	80

YIELD CLASS 160

20	1085	29	3¼	87	875	—	—	215	4	1.49	320	5	—	111	1195	8.3	160	60	20
25	745	38	4	87	1270	40	—	340	4	1.65	560	15	—	151	2150	7.5	214	86	25
30	525	47	5	93	1855	290	10	220	4½	2.54	560	40	—	186	3295	6.7	238	110	30
35	393	56	6	100	2505	930	180	132	5	4.24	560	85	5	217	4505	5.7	244	129	35
40	310	64	7	107	3155	1920	680	83	5¾	6.79	560	190	30	243	5715	4.8	237	143	40
45	256	71	8	113	3745	2850	1490	54	6½	10.3	560	290	80	266	6865	4.0	217	153	45
50	220	77½	8¾	118	4260	3550	2310	36	7¼	14.0	510	340	140	284	7890	3.4	194	158	50
55	194	82½	9½	122	4735	4160	3080	26	8	18.3	485	330	175	300	8800	2.9	168	160	55
60	176	87	10¼	126	5140	4650	3720	18	8¾	20.9	375	310	195	314	9580	2.5	145	160	60
65	162	90½	10¾	130	5470	5050	4240	14	9¼	24.6	335	290	205	325	10245	2.2	126	158	65
70	152	93½	11¼	133	5765	5380	10												—
75	144	96	11¾	136	6020	5660	5010	8	10¼	31.5	270	240	195	344	11360	1.6	97	151	75
80	137	98½	12	138	6230	5900	5310	7	10½	31.5	245	220	185	352	11815	1.4	85	147	80

Appendix II (c) (ii)

DOUGLAS FIR

NORMAL YIELD TABLE: YIELD CLASS 200

Age	MAIN CROP After Thinning							Yield From THINNINGS						TOTAL Production		INCREMENT			Age
	Number of Trees	Top Height feet	Mean BHQG ins.	Basal Area sq. ft. q. g.	Volume (h. ft.) to top diameter o.b. of 3 inches	7 inches	9 inches	Number of Trees	Mean BHQG ins.	Av. Vol. per Tree h. ft.	Volume (h. ft.) to top diameter o.b. of 3 inches	7 inches	9 inches	Basal Area sq. ft. q. g.	Volume to 3 inches h. ft.	C.A.I. Basal area	C.A.I. Volume to 3 inches	M.A.I. Volume to 3 inches	
15	870	33	3¼	65	830	—	—	430	3¼	0.47	200	—	—	97	1030	8.0	197	69	15
20	507	45½	4½	67	1255	80	—	363	3¾	1.93	700	15	—	137	2155	8.1	247	108	20
25	315	58	6	76	1895	630	100	192	4¾	3.65	700	80	—	177	3495	7.8	282	140	25
30	217	69	7½	87	2630	1840	830	98	6¼	7.18	700	300	70	214	4930	7.0	285	164	30
35	164	78½	9¼	97	3320	2880	2000	53	8	13.3	700	510	255	247	6320	6.1	270	181	35
40	132	86½	10¾	106	3925	3620	3040	32	9½	21.9	700	610	450	276	7625	5.4	252	190	40
45	111	93	12	114	4435	4210	3800	21	10¾	32.6	700	645	545	301	8835	4.7	233	196	45
50	97	98½	13¼	121	4940	4760	4450	14	12¼	42.8	620	590	535	323	9960	4.2	216	199	50
55	87	103½	14½	128	5445	5290	5000	10	13½	53.7	530	510	480	343	10995	3.7	198	200	55
60	80	108	15½	134	5925	5780	5510	7	14¾	65.9	470	460	435	361	11945	3.3	181	199	60
65	74	111½	16½	140	6355	6220	5960	6	15¾	77.0	430	420	400	376	12805	2.9	163	197	65
70	70	114½	17¼	145	6735	6600	6350	4	16½	88.4	395	390	370	389	13580	2.5	146	194	70
75	66	117½	18	149	7055	6930	6690	4	17¼	99.2	370	365	350	401	14270	2.2	129	190	75
80	63	120	18½	152	7305	7190	6960	3	18	109.0	355	350	335	411	14875	1.9	113	186	80

YIELD CLASS 180

15	990	30	3	65	740	—	—	310	3¼	0.26	80	—	—	88	820	7.4	168	55	15
20	593	42	4	66	1105	30	—	397	3¾	1.59	630	—	—	126	1815	7.6	223	91	20
25	375	54	5¼	74	1680	350	30	218	4½	2.89	630	40	—	164	3020	7.4	251	121	25
30	257	64½	7	84	2335	1300	430	118	5¾	5.36	630	160	20	199	4305	6.6	257	143	30
35	193	73½	8½	93	2975	2370	1350	64	7	9.92	630	375	130	230	5575	5.8	248	159	35
40	154	81	9¾	101	3550	3150	2390	39	8½	16.2	630	500	300	257	6780	5.1	233	169	40
45	129	87½	11	109	4045	3750	3230	25	9¾	24.7	630	560	425	281	7905	4.5	216	176	45
50	112	93½	12¼	116	4500	4280	3890	17	11	33.4	585	540	455	303	8945	4.0	199	179	50
55	100	98½	13¼	122	4960	4780	4450	12	12¼	40.9	495	475	430	322	9900	3.5	183	180	55
60	91	102½	14¼	128	5390	5230	4940	9	13¼	50.5	440	425	395	338	10770	3.1	166	179	60
65	84	106	15	133	5785	5630	5350	7	14¼	60.0	395	385	360	353	11560	2.7	149	178	65
70	79	109	15¾	138	6125	5980	5710	5	15	69.7	365	335	335	366	12265	2.4	133	175	70
75	75	112	16½	142	6410	6280	6010	4	15¾	78.0	340	330	315	377	12890	2.1	117	172	75
80	71	114	17¼	145	6640	6520	6250	4	16½	86.1	320	315	305	387	13440	1.8	103	168	80

YIELD CLASS 160

15	1240	27½	3	79	615	—	—	—	—	—	—	—	—	79	615	6.9	145	41	15
20	696	38½	3¾	66	985	—	—	544	3½	0.92	500	—	—	114	1485	7.1	196	74	20
25	454	49½	5¼	72	1485	160	—	242	4¼	2.31	560	20	—	150	2545	6.9	222	102	25
30	312	59½	6¼	81	2065	820	160	142	5	3.96	560	80	—	183	3685	6.2	231	123	30
35	231	68½	7½	89	2655	1810	770	81	6¼	6.96	560	220	45	212	4835	5.5	226	138	35
40	182	76	8¾	96	3195	2650	1710	49	7½	11.5	560	375	150	238	5935	4.9	213	148	40
45	151	82	10	103	3665	3270	2550	31	8¾	17.8	560	460	285	261	6965	4.3	198	155	45
50	130	87½	11	110	4070	3790	3240	21	9¾	25.6	550	475	370	282	7920	3.8	184	158	50
55	115	92½	12	116	4490	4270	3840	15	10¾	30.7	460	425	355	300	8800	3.4	168	160	55
60	104	97	13	122	4890	4690	4350	11	11¾	38.0	400	380	340	316	9600	3.0	152	160	60
65	96	100½	13¾	127	5250	5070	4760	8	13¾	45.5	360	345	315	330	10320	2.6	136	159	65
70	90	103½	14½	131	5565	5400	5110	6	13½	53.0	330	320	300	342	10965	2.3	121	157	70
75	85	106	15¼	134	5825	5670	5390	5	14¼	60.0	310	300	285	353	11535	2.0	106	154	75
80	81	108	15¾	137	6020	5870	5600	4	14¾	66.1	300	290	275	362	12030	1.7	91	150	80

APPENDIX III

Appendix III (a)

Forestry Commission

Form S.101

READY RECKONER FOR PROFITABILITY CALCULATION

(ENTRIES IN THE BODY OF THE TABLE ARE IN £ PER ACRE)

	3½% Allowance for:			5% Allowance for:			6½% Allowance for:		
	Standard	Extra haul to market	Difficult skidding	Standard	Extra haul to market	Difficult skidding	Standard	Extra haul to market	Difficult skidding
A. Discounted Revenue Yield Class									
60	90	5	10	20	—	5	5	—	—
80	150	5	10	45	—	5	15	—	5
100	220	5	15	75	5	5	30	—	5
120	300	5	15	105	5	10	45	5	5
140	380	10	20	135	5	10	60	5	5
160	460	10	25	165	5	15	85	5	5
180	550	10	30	200	5	15	110	5	5
200	660	10	40	235	5	20	130	5	10
220	790	15	45	275	5	25	155	5	10
240	920	15	50	320	10	30	185	5	10
260	1060	15	55	375	10	35	220	5	15
280	1200	15	60	440	10	40	255	5	15

	3½%	5%	6½%
B. Forest Maintenance Annual cost £ per acre	Capitalised	Capitalised	Capitalised
0.5	15	10	10
0.75	20	15	10
1.0	30	20	15
1.25	35	25	20
1.5	45	30	25
1.75	50	35	25
2.0	55	40	30

C. Roading Construction costs £ per mile	Total road density miles/sq. mile				Total road density miles/sq. mile				Total road density miles/sq. mile			
	2-3	5-6	6-7	8-10	2-3	4-5	6-7	8-10	2-3	4-5	6-7	8-10
500	—	5	5	5	—	—	—	5	—	—	5	5
1000	5	5	10	10	5	5	5	10	—	5	5	10
2000	5	10	15	25	5	10	15	20	5	10	10	15
3000	10	15	25	35	10	15	20	30	5	10	15	25
4000	15	25	35	45	10	20	25	35	10	15	20	30
5000	15	30	40	60	15	23	35	45	10	20	30	40
60000	20	35	50	70	15	30	40	55	15	25	35	45

This table may be applied as it stands when making comparisons between Yield Classes of the same species. When the Discounted Revenues (DR) of a number of species are being compared, adjustments should be made to DR. For the major conifers the adjustments to be made in the 5% column are as follows:

Pines S.S. G.F.—No adjustment required.

For the following species the DR should be REDUCED

N.S.   by 10%
N.F., R.C., L.C. by 15%
W.H.   by 20%

For the following species the DR should be INCREASED

D.F.   by 15%
Larches   by 40%

Appendix III(b)

Forestry Commission   **WORKED EXAMPLE**   Form S.102

CALCULATION OF N.D.R. FOR A PARTICULAR SITE TYPE

1. Working Plan Area HILL FARM: Site type: I VALLEY SIDES
   or Acquisition: ....
   Species: ..S.S......   Area (acres): 150......

	Rate of Interest	
	$3\frac{1}{2}\%$	5%
		— £ per acre —
A. Discounted Revenue		135
(i) Average yield class: 140		
(ii) Reduction on standard price		
(a) for transport haul: EXTRA 50 MILES		5
(b) for skidding conditions: NIL		-
(iii) Adjusted figure:		130
B. Discounted Expenditure		
(i) Establishment and brashing: £60		60
(ii) Maintenance annual cost: £1.0		20
(iii) Roading - cost per mile: £3000		15
miles per square mile: 4		95
(iv) Total		
C. N.D.R. = A(iii) - B(iv)		35
D. Discounted Investment		88
B(i) + (iii) + $\frac{2}{3}$ (ii)		
E. N.D.R. per £100 = $\frac{C \times 100}{D}$		40

Note: In acquisition appraisals ignore D and E.
Rates of interest of $3\frac{1}{2}\%$ and $6\frac{1}{2}\%$ are to be used only for
acquisition calculations and not for working plan purposes.

Date: ...16 January '64....

Compiled by: ...A. Smith.........

District Officer

AUTHOR'S NOTE: The Form also allows for a $6\frac{1}{2}\%$ rate of interest, not shown here.

Form S.104

FORESTRY COMMISSION

Appendix III (c)

WORKED EXAMPLE

CALCULATION OF OPTIMUM REPLACEMENT DATE

A. CALCULATION OF DISCOUNTED REVENUE FROM EXISTING CROP

Years from Start	0 Fell*	0 Thin	10 Fell*	10 Thin	20 Fell*	20 Thin	30 Fell*	30 Thin	40 Fell*	40 Thin	Remarks
(1)	(2)		(3)		(4)		(5)		(6)		(7)
A (i) Volume Yield / H. ft. per acre	2000	500	2300	600	2400	700	2500	600			Price of fellings (Year 0) low because stand unbrashed
A (ii) B.H.Q.O. inches	6	4⅞	7	5⅛	7½	6¾	8¾	7¾			
A (iii) Price	1s 6d		2s 0d	1s 4d	2s 3d	1s 8d	2s 5d	1s 11d			
A (iv) Value Yield = (i)×(iii)	150	25	230	40	270	58	302	58			
A (v) Discounting factor, 3½%	1.000		.709		.503		.356		.253		
A (vi) Discounted revenue £ per acre=(iv)×(v)	150	25	163	28	136	29	107	21			
A (vii) Insert final value for DR to present day in column appropriate to replacement date	150		163+25= 188		136+25 +28= 189		107+25 28+29= 189				

* Fell to exclude thinning production of the period.

B. CALCULATION OF DISCOUNTED REVENUE FROM SUCCESSOR CROP, NET OF DISCOUNTED EXPENDITURES VARYING WITH DATE OF REPLACEMENT

Replace at		£ per acre				Remarks
	0	10	20	30	40	
(1)	(2)	(3)	(4)	(5)	(6)	(7)
B (i) DR successor crop £135 discounted = DR × discounted factors A (v)	135	96	69	48		
B (ii) Discounting factor 5% for costs.	1.000	.614	.377	.231	.142	Brashing necessary before thinning area already roaded
B (iii) Cost of silvicultural operations £15 existing crop, discounted to present day and entered in appropriate column.	—	15	15	15		
B (iv) Cost of roads (£) similarly treated.	—	—	—	—		
B (v) Establishment plus brashing cost (£60) for successor crop similarly treated	60	37	23	14		
B (vi) Sum of discounted expenditure (= DR) = B(iii)+(iv)+(v)	60	52	38	29	19	
B (vii) DR successor not of DR = B(i) − B(vi)	75	44	30	19		

C. CALCULATION OF OVERALL NET DISCOUNTED REVENUE

Replace at	£ per acre				
	0	10	20	30	40
C (i) DR existing crop (from A(vii))	150	188	189	189	
C (ii) DR successor net of DE (from B(vii))	75	44	30	19	
C (iii) Overall NDR = C(i)+C(ii)	225	232	219	208	
Optimum replacement date		10 years			

DATE: 16 Jan. '64
COMPILED BY: A. Non

INDEX

Acland committee, 333
Adjustment period, 62, 189, 192, 198, 211–12
Administration, 132, 273, 348
Aerial photography and survey, 142, 162, 349
Affectation bleu, 130, 202
    unique, 130
Age, average, 173
    class, 94–96, 145, 170–3, 188
    exploitable, 64, 176, 179
    gradation, 93–100, 111, 188
    removal, 65, 77, 85–87, 111, 156–8, 184, 282
Albion, R. G., 334
Aldridge, F., 155
André, C. C., 189
André, Emil, 189
Annual plan of operations, 249
Arrangement of working plan, 253–60
Area, equiproductive (reduced), 55, 126, 171, 176, 319
    regeneration, 130, 201, 203, 228, 324
    statement, 301–2
    tending (and thinning), 126, 130, 146, 179, 203
    yield, 54–55, 170–88
Austrian cameral valuation, 188, 320
    formula, 189–93, 223

Balance of payments, 40, 41, 81
Balance sheet, 277, 294
Beat, forester's (sub-manager's), 132
Bellême, forest of, 323
Benefits, indirect (intangible), 18, 20, 31–35, 76, 112, 274, 343
    material (tangible), 20, 31, 35–36, 164, 274, 340–3
Bestandwirtschaft, 185, 332
Biblical references, 308
Binns, W. O., 163
Biolley, H. E., 103, 108, 213, 327, 330
Blendersaumschlag, 329
Block, forest, 114
    periodic, 125–30, 175–84, 322–4
    floating, 179
    permanent, 125–6, 176–8
    regeneration, 125–6, 128, 174, 178, 180–3

revocable, 127, 178
    single, 127–30, 179–84
    working, 114, 124
Bollem, 319
Bourne, R., 254
Bradley, R. T., 108, 291
Brandis, D., 335
Brandis method, 208
Brémontier, M., 325
Bruce, D., 142
Budget, 300

Cajander, A. K., 143
Cameral valuation, Austrian, 188, 320
Canute, King, 309
Chablis, 203
Chapman, H. H., 143, 147, 192
Charlemagne, 309
Charles V, 314
Class, age, 94–96, 170–3, 188, 273
    development, 145–6, 184–7
    size, 102–6, 207–10, 214–17, 273
    treatment, 145–6, 184–7, 303
    yield (quality), 99, 103, 158, 291, 293, 303
Clear cutting (felling), 53, 64, 92–3, 153, 325–7, 344–8
Climate, 223, 264
Coefficient of diminution, 105–6
Colbert, 316
Compartment, 114, 117–19, 129, 158, 305, 353
    history (register), 289–90, 301, 305
Control forms (of prescriptions), 252, 285–6
Control, method of, 103, 213–25, 290, 331
Conversion, 121, 324–5, 338
Conway, M. J., 150
Cotta, H. von, 53, 91
Cotta's formula, 174, 177, 205
Coulon, M. de, 108
Couvet, 108, 218–22
Current annual increment, 68–69, 98–101, 190
Cutting (felling) cycle, 131, 208–10
    (felling) series, 122–4, 279–80
    motives, 165
    section, 124
Cycle, felling (or cutting), 131, 208–10
    thinning, 131, 273